Culture on the Brink

Ideologies of Technology

Dia Center for the Arts

Discussions in Contemporary Culture

Number 9

Culture on the Brink

Ideologies of Technology

Edited by Gretchen Bender + Timothy Druckrey

Bay Press - Seattle - 1994

Printed in the United States of America
First Printing 1994

Library of Congress Number 89-650815
ISSN 1047-6806
ISBN 0-941920-28-3

Bay Press
115 West Denny Way
Seattle, WA 98119
206.284.1218 (fax)

Designed by Bethany Johns Design
Edited by Ted Byfield

"The Dream of the Human Genome" from *Biology as Ideology: The Doctrine of DNA*
by R.C. Lewontin. Reprinted with permission from *The New York Review of Books*.
Copyright © 1994 Nyrev, Inc.

"Stories from the Nerve Bible" from *Stories from the Nerve Bible* by Laurie Anderson.
Copyright © 1994 by Laurie Anderson. Reprinted by permission of Harper Collins
Publishers, Inc.

"Virtual Reality as the Completion of the Enlightenment Project" by Simon Penny
is a revised version of an essay, which will also appear in *Virtual Reality: Case Histories*,
edited by Carl Eugene Loeffler (New York: Van Nostrand, 1994).

"Give Me a (Break) Beat! Sampling and Repetition in Rap Production" by Tricia Rose
has also appeared in *Black Noise: Rap Music and Black Culture in Contemporary America*
(Hanover, NH: Wesleyan University Press, 1994).

Cover image: Gretchen Bender, *Untitled* (1991)

Page 13: Still from Nam June Paik, *Lake Placid* (1980). Photo: Marita Sturken
Courtesy Electronic Arts Intermix
Page 83: AIDS virus budding from plasma membrane. Copyright © Photo Researchers, Inc.
Page 147: Satellite image of Chicago, Illinois. Copyright © CNES/Spot Image/Photo
Researchers, Inc.
Page 205: Courtesy Gretchen Bender
Page 265: Courtesy Photo Researchers, Inc. Copyright © 1991 Lowell Georgia

Contents

I. Ideologies of Technology

II. Technology and the Body: The Constitution of Identity

III. Information, Artificiality, and Science

IV. Technology, Art, and Cultural Transformation

V. Technology and the New World Order

Dedicated to Jim Pomeroy

Less than a week before the *Ideologies of Technology* conference, Jim Pomeroy, a challenging and wonderful artist, unexpectedly died in Texas. The paper he planned to present at the conference was pure Pomeroy. As a study of the Biosphere 2 project in Arizona, it ranged over architecture, performance art, and new age aesthetics, with references to Disney's Epcot Center, Paxton's Crystal Palace, Archigram's fantasy megastructures, Galleria Shopping Malls, the Mercury Astronauts, the Mouseketeers, and American Gladiators. The title, "Performance and Architecture in New Age Spectacle (or People Who Live in Glass Houses Shouldn't Pass Gas)," was the usual groan-producing Pomerian pun, science bass-ackwards, performance in architecture, and new age spectacle. His own spectacle, the many images he had gathered to accompany his talk, were taken from tourist slides sold in the Biosphere 2 gift shop or shot with the impressive array of cameras he managed to get past the vigilant Biospherian security team. Jim was our foremost tinkerer, Pied Piper, and popular mechanic. Like Jim, we can all aspire to be popular scientists and popular artists, too.

Constance Penley

A Note on the Series

Since 1987, Dia has presented a series of symposia and accompanying publications to bring cogent voices to critical issues in contemporary cultural discourse. This book, the ninth in the "Discussions in Contemporary Culture" series, is the timely conception of Gretchen Bender and Timothy Druckrey. Examining the shifting roles of technologies in diverse public and private spheres, such as work, leisure, art, and war, the texts included here were presented by the participants of a three-day conference held at Dia in Spring 1992.

We would like to extend our warmest thanks to Gretchen Bender and Tim Druckrey for their careful structuring of the *Ideologies of Technology* conference, and to the participants for their contributions and their time spent editing and rewriting papers for publication. Karen Kelly, Director of Publications and Program Coordinator, took the lead role at Dia in shaping the symposium and this publication and has committed her great energy and intelligence to this project since 1991. She was assisted primarily by Franklin Sirmans. Lynne Cooke, Dia's Curator, has also given generously of her time and wise counsel to the organization of *Ideologies of Technology*. For their good humor and resilience during the frantic weeks of organization prior to the conference, we gratefully acknowledge Sara Schnittjer Tucker, Laura Fields, Steven Evans, Michelle Marozik, and the staff at our 22nd Street location.

We are indebted to Ted Byfield for his editorial skills during compilation of the essays for this book. It has also been a pleasure to work for the first time with Sally Brunsman and Kimberly Barnett at Bay Press. The design of this volume by Bethany Johns, and her assistants Georgie Stout and Marion Delhees, is, as always, deeply appreciated. We also thank Anastasia Aukeman, Kristin M. Jones, and Phil Mariani.

Public programs at Dia are supported in part by a generous grant from the Lila Acheson Wallace Theatre Fund at Community Funds, Inc.

Charles Wright
Executive Director, Dia Center for the Arts

Introduction

Timothy Druckrey

Technology forms the core of the cultural transformations that are generating startling changes in virtually every cultural and political activity. For almost half a century, maturing technologies have introduced radically reconfigured forms of representation, a refunctioned industrial base, a reinvented communication system, and a medical system readying itself to find and treat disease as a form of genetic defect instead of a manifestation of complex historical and cultural history in which the emergence and constitution of illness is located. It is now common to think of postindustrialism and postmodernism as related to the multinationalization of technology—a New World Order reliant on instantaneous digital communication and essentialist assumptions of a global village finally coming to fruition. With the legitimation of cultural technologies widely seen as contributing to social progress emerges a set of ideologies whose function serves as the rationale for stability. Yet the concept of a singular—one even might say modernist—ideological structure no longer serves to rationalize cultural change, nor to sustain the unity of conditions of difference. Rather than one technology, many; rather than one ideology, many. What is emerging is a discourse of ideologies forming within a distributed set of technologies.

Even though technology's effects were pivotal to its formation, its formative structure has been excluded from early discussions of postmodern culture. This lapse suggests an equivocal stance concerning technology itself. Modernity matured in the economies of both capitalism and science; indeed, the culminating merger of these two systems has formed the foundation of culture of the past two centuries. If the shift from modernism was, in some ways, precipitated by the triumph of technology over speculative science, then the assessment of technology within postmodern culture is premised on the triumph of technology

[1]

over experience. Productive technologies altered the balance between progress and nature. But technologies are being encouraged that alter the terms of our relationship with the far more subtle effects of our inherited genetic history, our psychological history, and our knowledge of the world. Technology—in some ways, one can focus on the technologies of representation—mediates and, increasingly, simulates the experiential in ways that complicate and confound many of our notions of reality. In the introduction to his novel *Crash* (1973), J. G. Ballard writes that "we live already in a world ruled by fictions of every kind. The fiction is already here and the role of the writer is to invent the reality."[1]

Our technologies and our fictions are converging. Promises of dizzying access to information, communication, and electronic communities suggest the kind of conjecture that fills science fiction. Virtual futures propose rather than substantiate; instead of necessity they offer only possibilities, a kind of imaginary future in which we will participate as agents of techno-progress and consumption. The boundaries that so long characterized the distinctions between the speculative and the instrumental have been eroded by a merging of research, development, and marketing, a reaction characterized by the commodification of technology as immediate and intrinsically productive—too often without thorough reflection. Enveloped in the demands of immediate production and consumption, technologies become convincing mostly because they seem to function invisibly. The privatization of the technology market and the lapse of other short-term goals positions technology as the essential media of postmodern culture, a pastiche of instrumentality and speculation. As the globalization of high technology is positioned as both a panacea and the final blow to industrialization, high technology too has been assimilated into creative production. Never before has there been such an integrated transformation of culture.

Technology, in guises touted as "user-friendly" or as "radically sophisticated," has become the amorphous foundation of a new social order. The so-called information superhighway is supposed to provide the circulatory system of a virtual communication order; smart technologies are supposed to monitor and learn as they are used (no doubt providing comforting surveillance); nanotechnology promises to create molecular machinery to repair the body and to provide a kind of industrial immune system. Of course, the promise of biogenetics abounds in claims for genetic mapping as a revolution in health, reproduction, and treatment. Already, the future is being mapped in terms of the dynamics of possible technologies rather than in terms of appropriate or responsible ones. But it is essential to understand that evolving electronic technologies are already incorporated into even seemingly passive devices, such as the VCR or the telephone, which have radically altered everyday social relations.

Perception, memory, history, politics, identity, and experience are now mediated through technology in ways that outdistance simple economic or historical analysis. Indeed, technology pervades the present not simply as a mode of participation but as an operative principle. Beneath the facades of ownership through consumption and conceptualization through use, technology subsumes experience. The relationships between technology and knowledge, class, scarcity, and competition can no longer be framed in strictly economic terms; they have encompassed the individual. In forms that assail the boundaries between understanding and certainty, technology is operational not merely in the formation of ideologies but also in the practices of everyday life. Immersed within electronic environments, the experience of the individual conforms to the logic of the technological or is characterized as reactionary or sentimental. Donna Haraway's work forcefully identifies this phenomenon in her seminal essay "A Manifesto for Cyborgs." She writes, "Late twentieth-century machines have made thoroughly ambiguous the difference between the natural and the artificial, mind and body, self-developing and externally designed ... Our machines are disturbingly lively, and we ourselves frighteningly inert."[2]

[3]

While the individual effects of technology are clothed in both fascination and desire, a reliance on new forms of technology assaults individual autonomy: technology concentrates power as it deconcentrates and insubstantiates the individual. The effects are evident in a broad range of areas, and can be seen most pertinently in the spinoff forms of surveillance that digital technology introduces—monitored keystrokes, monitored messages, banks of genetic data, fully recorded message systems, caller identification, and so on—already widely utilized as elements of the benefits of technologization. The electronic environment is ripe for data collection on an unprecedented level, even though the dispersal of these technologies is often marketed as convenience. But the consequences of vast databanks threaten an already fragile concept of freedom. Deconstructing technology, demonstrating that its effects can rely on superficial notions of progress, is a daunting task. As culture relies on the pervasiveness of the technological in ways that are revealing and intricate, we are often lured into hazardous complacency. Too long exempt from critical scrutiny, the issue of electronic technology stands as the most demanding denominator of current culture.

It is no less obvious that technology is interlaced within the structures of existence. Surrounding us in what Peter Weibel has called an "intelligent ambience," we assume that technologies will monitor and provide an active presence in life. Though often paralleled with the adaptivity of the biological mechanism, technology is not organic. Yet the anthropomorphization of technological accessories suggests strategies that humanize their uses. Personal digital assistants, portable phones, and laptop computing all contribute to the assimilation of high technologies within normal activity. Nevertheless, the rationality of the connection of technological development with that of cultural development is pervasive and convincing. Adaptation and conformity are not synonymous, even if they seem so in the ambience of technology. The Darwinian mechanism is negated as an entity reliant on genotypical events, but the logic of technology is developing more and more within frameworks of biological determinism. Adaptability is measured no longer in terms of survivability but in compliance with the conditions of information: biotechnic standards. Technology outdistances the "natural" and substitutes the domination of nature. The entire rationale of technology suspends the instrumentality of nature and serves to demonstrate the instrumentality of technology. Accomplishment, experience, and authenticity play little role in the inexorable development of the mechanisms of technology. And what more cogent metaphor for the blurring of the boundary between technology and biology than genetic programming, the reduction of biological change to computed algorithms, artificial life programs, a bank of genetic codes? The Human Genome Project will attempt to configure the most extensive collection of genetic data in history. Its goal will reshape the evolution of the biological mechanism. Encoding rather than identity could become the signifier of the self, an informatics of domination made possible only by the power of the computer.

As a fundamental element in the experience of culture, technology serves in several forms. Entertainment, communication, work—the web of technologized experiences, in Martin Heidegger's terminology, "enframes" the way things are done. It reinforces the idea that compulsions of development are bound unalterably to technical growth. Technology is assimilated into popular culture in ways that reinforce its authority but mask the tactic of domination. Games, news, film, music, photography, and so on, often serve as the testing ground for technology, offsetting research and development costs often as marketing strategy. Advanced technology is integrated as novelty and as necessity. The exploitation of culture seems no longer to be linked directly with labor but with a more problematic connection—desire. To be "disconnected" stands as a form of impotence, of lack, of alienation.

How technology and representation establish desire as a denominator is an issue rooted in the growth of the "consciousness industry." In an era when the structures of consciousness are themselves under siege by the adherents of intelligence models that tout their artificiality (one might identify the emerging model as the "cognition industry") resolving the issues of knowledge and the relationship between cognition and representation is pivotal. How far this seems from the culture industry of Theodor Adorno and Max Horkheimer! Artificial intelligence, conjoined with information technology and simulation modeling, could supplant the relation between knowing and consuming. Indeed, as the distinction between research, information, conceptualization, and expression are determined by computerized potentials that funnel expectation and thwart imagination, knowledge may become a form of consumption. This technological unconscious is a parallel to Fredric Jameson's "political unconscious."[3]

If the political unconscious represents a fundamental grasp of the "ideology of form," the relationship of necessity to expression, then a critique of the technological unconscious requires a deeper consideration not only of culture, society, and history, but also of the psychological and ideological effects of form immersed in the illusions of artificiality, simulation, immateriality, and the virtually "real." In the shift from a discourse concerned with the material critique of representation—of gender, of sexual politics, of the institution of narrative—there is a merging of discourse of the postmaterial, the cybernetic, the virtual. How we root this speculative discursive assessment within pragmatic intent will be something of a challenge. Jameson's remark that "no society has ever 'embodied' a mode of production in any pure state"[4] is, in some ways, being altered by the intrusion of "productivity" into the biological roots of behavior or consciousness at the same time that technoculture offers itself precisely as the embodiment of production. The discourses of technology now encompass both speculation and politics. The shift from industrial culture to media culture to information culture to technoculture represents a phase shift of daunting proportions.

The decade of the 1990s is one that must absorb and contend with the sea changes wrought by technoculture. The task for social criticism is daunting, for not only must the structural transformations both generating and generated by new technologies be identified, but critics themselves must become adept in the language and use of the very technologies under scrutiny.

"Since the end of the nineteenth century the other developmental tendency of advanced capitalism has become increasingly momentous: the scientization of technology."[5] Jürgen Habermas's essay "Technology and Science as 'Ideology'" set the rationale of power at the heart of the development of culture. Though the essay correctly links legitimization with the market, it does not fully realize either the potential of technology merged with science, or the staggering effects of a culture encompassed by representations legitimized by the discourses of capital, science, media, photography, and computers. In order to approach the significance of technologies that will supplant photographic representations, representation must again be deconstructed. Nowhere is the evidence of this deconstruction more problematically signified than in the computer.

In less than a decade, culture has undergone a fundamental shift as the computer embraces an increasing range of tasks. Yet the administrative efficacy of the machine is eclipsed as it begins to assimilate representation itself. Imaging systems have come to echo the forms of representation that so dominate this culture. Digital-imaging technologies now routinely use computed sequences to dramatize their content. Virtual commercials, animation, and "morphing" have become indispensable components in making imaging relevant to a generation of users comfortable with the electronic illusory. At the same time, video, film, and principally photography are being challenged to hold their authority against visual modeling systems that are emerging to eclipse its forms. These imaging models have their practical roots in the photographic, but their conceptual schemes lie in simulation. As representation and technology converge, a crisis emerges. The configuration of data used to construct an electronic image is epistemological—based on a knowledge of how the structure of perception can be rendered— while the photographic recording is phenomenological. This is a fundamental reconfiguration of the model of what constitutes an image, one that raises the issue of whether an image can serve as objective information. As such, the images sever the link that photographic representation presumed for so long. While this might be perceived as constructive, it also denotes a form of detachment that fragments experience even more deeply than the media environment that preceded it. As Paul Virilio writes, we will live within a "logistics" of perception. Whatever the material relationships that linked perception, history and the future are at stake.

New technologies supplant and disenfranchise precisely when they seem friendly. At a recent lecture, one member of MIT's Media Lab blithely said that we are going to make a transition from computers you're in front of to computers you're inside of where the computer becomes so present that you almost wear it. This would seem to be almost nonsensical if the issue of "virtual reality" were not already implementing the idea. In virtual reality, the senses are not so much deceived as superseded. VR represents a series of developments in the human computer interface that have quickly reached into the public's imagination. In VR, visual experience is generated by powerful computers that track the movement of the head and body and render, in pseudo–real time, a dimensional "presence" within a virtual environment. Use of the prosthetic devices of the virtual system—head-mounted displays, data gloves, and body suits—encase the body in fiber-optic cabling. Movement is computed and a representation of the body is displayed within the scene. Immersion, the absorption of the senses within the system, represents a triumph of technology and a challenge to structures of representation. "Eyes and mouth," writes Friedrich Kittler,

> long accepted as the last refuge of intimacy, will be integrated in a feedback logic which is not ours, but that of the machine. The affect of VR would not be so unsettling if we did not realize that we are being observed by something that we cannot acknowledge as subject or persona in the traditional sense, and which nonetheless constantly demonstrates that it sees us without revealing itself.[6]

What is so fascinating about the issues of immersion and virtual reality is the way in which it substantiates the problem of representation itself. Slavoj Žižek writes:

> The ultimate lesson of the virtual reality is the virtualization of the very "true" reality. By the mirage of "VR," the "true" reality itself is posited as a semblance of itself, as a pure symbolic construct. The fact that "the computer doesn't think" means that the price for our access to "reality" is also that something must remain unthought.[7]

Participatory "worlds," tactile feedback, and the liberation of the body from the constraints of material imagination pinpoint a significant problem in the depiction of an authentic representation of a reality in any substantive sense. As Jacques Lacan expressed it, the Real is that which is unrepresentable. As the system of technology expands to dominate the regulation of the external world, it also contracts and increasingly penetrates the internal world. The body is unquestionably the next frontier—the body, and then cognition.

The penetration of technology within the body, and the socialization of simulated realities are more than signifiers of technological progress. They mark a radical transformation of the social order in which knowledge is linked with ideology, biology, or identity in terms of a technological imperative not fundamentally connected with necessity. The consequence of genetic engineering (or, perhaps more appropriately, genetic therapies), of patented life forms, and of radicalized technomedicine and technopsychology, are among some of the more sweeping ethical issues of our time. As technological research is privatized and perhaps excluded from government oversight, the funding, projects, targets, ethics, and manipulation of development become suspect, not because government supervision is more effective (as is clear from its appalling record on AIDS research), but because it is assailable as part of a more visible history. Basic science succumbs to applied technology. The corruption of science, fueled by the economics of technology, could herald a culture ornamented with gadgets but devoid of public accountability. Rarely in the debate about technology does the voice of the citizen emerge. Specialized discourse, reduced to elemental soundbites, determines policy to an unprecedented degree. A technologically illiterate culture, inebriated with the spinoffs of high technology, hardly pauses to address the issue of the impact this environment will have.

Information has become the lubricant for a swiftly emerging social structure that is wholly dependent on the potential, malleability, and exchangeability of data. Evident in the recent merger-mania between broadcast, telephone, and computing companies, the near future of basic communication, broadcasting, and the consumer have taken on a frenzied urgency. "Have you ever tucked in your kid from a phone booth? Have you ever read a book from two thousand miles away? Have you ever faxed someone from the beach?" asks a recent campaign from AT&T. Their reply: "You will." Selling the future after "we're all connected" is becoming a trend in current advertising. Marketed as intimate and essential, communication would seem to eradicate the need for anything other than telepresence—an obvious benefit of postmodernization. The already unfolding issue of intellectual property rights reveals the vulnerability of the individual in a system captivated by innovation. The established relationship between product development as the resolution of necessity and rampant development in search of necessity is a telling metaphor for a culture consumed with desire. It was, for example, not surprising that the American media covered the fall of the Berlin Wall as if it were the prelude to a shopping spree. The images of gleeful, formerly "oppressed" citizens climbing back into East Germany shouldering televisions, radios, and electronics of all sorts makes it plain that the desirability of democracy is consumption. Instead of an old-fashioned land rush, there has to be a technology rush, followed by the corporate troops swarming in under the banner of democracy to loot and liberate the exhausted social system from its justifiable fear of capitalism.

Electronic cities, virtual communities, information superhighways, cyberfactories, and *cyberias* are among terms already making their way into an evolving cultural technolanguage. As language adjusts to technology, the limits of representation become clearer. The specificity of the languages of programming are without metaphor, without ambiguity: an instruction set describing logical possibilities is not the same as a language whose effectiveness and development is premised on evolving significations. If there is a denominator for the language of technology it is the binary system, the groundwork of the electronic communication. Rooted in the communications technology of the electronic, one can say without doubt that the so-called New World Order will be digital. The rationality of the culture of the digital is wrought with discursive problems. Grounded in models of rationality whose legitimacy is questionable, the development of technologies of representation are reflexive merely as models that are already considered complete. A discursive, even a deconstructive relationship with them is difficult. Can one, for example, deconstruct the programming codes of rendering in social terms? Yet the effects, the history, and the social logic of digital coding demand this analysis. What emerges in the linking of technology, representation, and a cultural space increasingly wired into the global network is a technocratic ideal premised in democratic principle.

The effects of technology demand great scrutiny, and a creative series of responses have surfaced. Alongside rapid developments in the fields of multimedia and digital video have come a number of artists, writers, musicians, and designers willing to explore the potential of computer technologies, sampling, and hypertextual fiction as media whose imaginative potential poses—or perhaps opposes—questions about the role of the artist in technoculture. Using digital editing of photography and video, initial steps are being made by artists to develop hypertextual narratives whose plots are not merely described but are "linked" to tentative projects exploring virtual reality as an imaginative space (rather than an absorbing video game). As interactive art and media implicate the participant in experiential media, offering feedback as an essential component of artistic production, they pose new questions about the issues of commodification. Sampling has emerged as an issue both in creativity and in intellectual property rights. In an information order, "ownership" becomes an ambiguous issue that challenges the legal establishment to reconsider copyright, fair use, and what exactly constitutes property. Professional scientific organizations like Computer Professionals for Social Responsibility and the Electronic Frontier Foundation have begun to intervene in exposing the political and ethical issues that develop simultaneously with new technologies. Similarly, "communities" have begun to spring up along the information superhighway whose distributed interests and discussions could serve as the beginning of networked discourse and information sharing. The issue of empowerment reemerges as these interest groups organize.

Culture on the Brink: Ideologies of Technology is not an account of the frenzy of invention but a critical forum for an approach to the question of technology as it pertains to the culture of experience. As technology is assimilated into the structures of the everyday, its ubiquity becomes invisible and necessary. A generation of smart technologies—everything from smart refrigerators to smart cars—will learn to anticipate and to act autonomously based on discernible patterns. Invisible technologies will surveil the intricacies of everything we do. We now assume that technology is comfortably omnipresent. The goal of this project is to frame a critique of technological reason, to deconstruct the mythology that technology is a panacea. In directly implicating technology within the forms of everyday life, and by demonstrating that disaster will be the likely consequence if the public remains uninformed, we have attempted to position technology as a cultural form as prone to manipulation as public opinion. Its effects, though, are potentially more insidious and its privatization more alarming. The essays here are not so much diagnostic; rather they confront the use of technology as alternative practice. Without denying the impact of the cultural attention paid to the technological, it would prove useless to avoid politicizing activity by simultaneously employing and depleting the efficacy of the technological illusion.

1. J. G. Ballard, *Crash* (London: Paladin Books, 1990), p. 8.
2. Donna J. Haraway, "A Manifesto for Cyborgs: Science, Technology, and Socialist-Feminism in the Late Twentieth Century," in *Simians, Cyborgs, and Women: The Reinvention of Nature* (New York: Routledge, 1991), p. 152.
3. Fredric Jameson, *The Political Unconscious: Narrative as a Socially Symbolic Act* (Ithaca, N.Y.: Cornell University Press, 1981).
4. Ibid., p. 97.
5. Jürgen Habermas, "Technology and Science as 'Ideology,'" in *Jürgen Habermas on Society and Politics: A Reader*, ed. Steven Seidman (Boston: Beacon Press, 1989), p. 252.
6. Friedrich Kittler, "Conversation," in *On Justifying the Hypothetical Nature of Art and the Non-Identicality within the Object World* (Cologne: Galerie Tanja, 1992), p. 135.
7. Slavoj Žižek, "From Virtual Reality to the Virtualization of Reality," in ibid., p. 135.

I

Ideologies of Technology

Technology and the Future of Work

Stanley Aronowitz

No doubt the main ideology of modern technologies is that *virtually* all of our problems—ethical, economic, political—are subject to technical solutions. For example, cybernetics is presented as the genie that, once liberated from the bottle, will possess the powers of an old-time elixir: it can cure virtually any disease (except death) or even change the genetic makeup of the human species, eliminating those characteristics "we" agree are unwanted, and it can solve most economic ills by spurring growth. (Indeed, an American, Robert Solow, won a Nobel Prize in economics for this amazing insight.) If any employer's profits are squeezed by skyrocketing costs, install a computer: watch heads roll and the bottom line soar. A new science, teledildonics, is being born, which may offer virtual sex (at a distance, of course) to the weary hacker—an innovation that, together with cybernetically charged virtual toys, promises to cool us out.

We live in a postcritical period; not only practical but also intellectual luddism seems irreversibly weakened. The merger of the intellect with technology signals the demise of the ancient naysayer. This negative dialectic yields to the power of positive thinking: while philosophers and social critics ask what implications this wraparound technology has for such old metaphysical questions as the integrity of the "self," the celebrants of (take your pick) cyberspace, cyberpunk, virtual reality, hyperreality, and its literary form, hypertexts, refuse the question as a modernist hangover. When not following the poststructuralist Nietzschean mantra about the "death" of the subject of which, in classical Anglo-American philosophy, the self is an outgrowth, the new technophiles gleefully announce that we are totally wired and it is entirely futile to cling to the old formulations of, say, the Frankfurt School, which speak of technological domination of the once autonomous subject.

[15]

Apocalyptic ruminations by philosophers and social theorists of the thorough commodification and technicization of social life are refuted but also mirrored by the strategic amnesia of the technophiles. Perhaps we need to comprehend what has occurred in Donna Haraway's openly ironic terms: humans have migrated to cyborg, "a hybrid of machine and organism, a creature of social reality as well as a creature of fiction."[1] Haraway calls our attention to borderlands, not only between reality and fiction and between machine and organism but perhaps more saliently between the biological and the social self which is congealed in our most ubiquitous production: the machine. Far from a happy conjunction, "the relation between organism and machine has been a border war. The stakes in the border war have been the territories of production, reproduction, and imagination."[2]

Whether employed as a powerful instrument in the corporate panopticon or as a way to facilitate the emergence of a democratic workplace in which *techné* signifies the unity of humans and nature rather than a technologically driven split, the machine is indeterminate from the perspective of the internal constitution of the technology. I suggest that many of the widely cited instances in which the introduction of the computer into the professional work formerly performed either by hand (as with engineering) or by conventional oral and printed texts (as with teaching) manage to elide the panopticon of power only by "forgetting" the context within which computer-mediated work is done.

The critique of science and technology that reemerged in the 1950s and 1960s hinged, in the final accounting, on the judgment, most forcefully articulated by the Frankfurt School, that reason under late capitalist conditions had become identical with instrumental rationality, whose basis is the capacity of scientifically based technology to facilitate a new social deal. The key element of the deal was that technology could "deliver the goods" to the underlying population in the West (and possibly elsewhere as well), in return for which culture surrendered its autonomy to the technological imperative. For Max Horkheimer, the new technology that came into existence and provided the material conditions for institutions of massified, electronically mediated culture signified nothing less than the eclipse of reason as a transcendent principle and, accordingly, the end of history. The consequence of technological domination was nothing less than the disappearance of historical agency, if by that term we mean both the will and the capacity of exploited and otherwise excluded social categories to participate in the polity as well as share control of society.

These powerful suggestions partly animated the massive student revolts of the 1960s and the "Third World" revolutions of the postwar era. Although they have outlived their moment of ideological power, they nevertheless retain much more than a marginal existence among important sectors of Western intellectuals to this day. However, in the past twenty years the technoculture that was once received as merely a brilliant and dystopian anticipation in social theory of what Daniel Bell and Alain Touraine would dub "postindustrial" society—a society in which social contradictions give way to conflicts that are subject to technologically based solutions—has become an ineluctable feature of everyday existence (conjoined, undoubtedly, with the ebbing of the most politicized manifestations of the resistance).

Technoculture has its prophets as well as its critics; like all other cultural spheres—education, sexuality, sports, art—technoculture realigns the discursive combat. One may no longer *derive* a position on the new computer-driven culture from familiar ideological premises. For example, within the new discursive boundaries, it is difficult to discern the differences on the basis of the traditional divisions between Marxists and liberals, since both argue (from different premises, to be sure) that the task of public policy is to free the new scientifically based technologies from the thrall of arcane social arrangements. In turn, critics of the scientific and technological revolution of our time have emerged from both conservative and radical camps, and their elegiac invocations of past civilization are remarkably similar.

[17]

Martin Heidegger's subtle questioning of the instrumental view—according to which technology is merely purposive activity subsumed under the characterization "means"—may be taken as a starting point for investigating the various claims concerning the new technologies. Recall that Heidegger brought us back to the Greek origin of the meaning of the term *techné* in order to free technology from its subsumption under instrumental reason. Where modern culture views technology as a regime of powerful tools by which human purposes may be served, particularly the domination of nature, *techné* signifies an uncovering, a way to the truth. Heidegger's point is that *techné* signified human activity itself rather than a "tool" of production and organization. In this reprise nature is not held at a distance as "other" but, as in ancient times, humans are *part of nature*. Thus, contrary to the practices of the past five centuries, nature is not perceived as an inert object. Its immanence includes us.

In contrast to *techné*, by which the object, nature, discloses itself, modern technology is an "ordering revealing," and technology becomes a process by which reality is teleologically enframed. This enframing blocks the truth of the world from "shining forth." In this regime, the world stands "in reserve" of human purpose, and it loses its own essence, at least provisionally. Of course, Heidegger's project is to restore to nature an autonomous existence from which *techné* then receives its rationale. The instrumental view, according to which technology is a means, does not, thereby, lose its power even if we restore *techné*; the history of the last twenty-five hundred years is irreversible. Rather, instrumental action becomes a local activity subordinate to its meaning as a way of revealing the being-of-the-world.

We can discern in Heidegger's discourse a not-too-veiled attack on the position of the Frankfurt School and other critics of modern technology. Against those for whom technology has *become* culture in late industrial societies, to the detriment of the autonomy of thought and being, Heidegger—despite his acknowledgment of the force of this critical view—wants to reconcile modern technology with culture by reviving the former's status as an end that extends beyond the category of domination. Echoing the early Marxist lament—that as machinery rendered human labor increasingly productive, labor is less able to control not only its product but also the process of labor—the Frankfurt School attempted to show that this dialectic strengthened domination. Specifically, to the extent that technological development signified the enlargement of human mastery over nature, it also created new mechanisms for human domination. However, by the mid-sixties, Herbert Marcuse had become ambivalent about this judgment. He returned to Marx's last meditation on technology, which saw technology's most subversive effect in its tendency to "liberate" labor from the production process. Automation would free labor for creative activity, some of which would be devoted to further reducing necessary labor.

Technoculture is a discourse of the identity of technology and culture. As we have seen, in its most sweeping expression, the cyborg, the distinction between humans and nature is called into question. While the organism and the machine may be historically, even analytically separate, Haraway's invocation of the metaphor of the border is only one version of a much larger movement in philosophy and culture theory. Thirty years ago, Louis Althusser's interrogation of the doctrine of humanism demonstrated the historicity of the pristine concept of "man." Invoking the figure of Spinoza, Gilles Deleuze's work seeks to free philosophy and culture theory from its idealist remnants, which, like Althusser, he finds in the Hegelian legacy. Deleuze has restored Spinoza's doctrine, which sunders the separation of spirit from "dead" matter in favor of immanence: all there is is the material world.

From this doctrine follows the materialist refusal of the Frankenstein/golem imagery invoked by technology's humanist critics. No longer critical, it accepts, even celebrates, computer-mediated technology as liberating. For example, in this postcritical technology we are informed of the wonders of cyberspace, where overworked professionals can play therapy against stress with innovations such as virtual reality. In cyberspace the old-fashioned "pressing of the flesh" yields to what Howard Rheingold calls teledildonics:

> The first fully functional teledildonics system will be a communication device, not a sex machine. You probably will not use erotic telepresence technology in order to have sexual experiences with machines. Thirty years from now, when portable telediddlers become ubiquitous, most people will use them to have sexual experiences with other people, at a distance, in combinations and configurations undreamed of by pre-cybernetic voluptuaries. Through a marriage of virtual reality tech-

nology and telecommunication networks, you will be able to reach out and touch someone—or an entire population—in ways humans have never before experienced. Or so the scenario goes.[3]

Rheingold tells us that "You can reach out your virtual hand, pick up a virtual block, and by running your fingers over the object, feel the virtual surfaces and edges, by means of the effectors that exert counter-forces against your skin."[4]

But, of course, the use of cybernetics for play or for producing nonlinear texts (hypertexts) are only a few of the uses of this technology. Equally, if not more to the point, cybernetics has revolutionized the workplace, biological science, and so on; it presents a whole new set of medical problems as well as possibilities.

Ironically, economic discourse has almost entirely disappeared from technophilic rhetorics. For those who herald cyberspace as the fulfillment of *homo ludens*, as the new sphere of the playful intellect, the disappearance of *homo faber* is considered an accomplished fact; postwork is taken as a given of the contemporary world. In the bargain, celebrants of the new technological revolution engage in massive historical amnesia or, worse, have chosen to refrain from commentary on the major practical consequence of cybernetics: the destruction of labor. It might be said here that those who would refuse to speak of this massively central function of the scientific-technological revolution should also keep silent about its playful aspects.

Put simply, cybernetics is the means by which labor is more and more removed from the industrial, commercial, and professional workplace. Its major function in our culture is to institutionalize unwork. Its effect is to raise the question "Why work?" in ways that were once part of the utopian discourse of technophiles.

Computer-Aided Design and Drafting (CADD), for example, has *virtually* eliminated the drafter in, and radically changed the character of, engineering. The articulation, by means of a computer, of design with execution (manufacturing) has increased productivity exponentially, wiped out whole occupations, and transformed the work from its tactile, craft character into a surveillance function.

Throughout the modern epoch, proponents of the scientific enlightenment have taken the position that the "idyllic" relations of precapitalist societies were impossibly repressive to the human spirit, and that capitalism, the evils of exploitation notwithstanding, had its one redeeming feature in the will to change the world—led by scientific and technological transformations. However, it is not excessive to argue that the computer, even more than its anterior concomitants (for example, steam power and electricity), provides an even more persuasive case for the claim, espoused by the newest technophiles, that finally, after nearly three centuries of rationalization, we can now envision the reintegration not only of work but also of humans with nature and with their own species. For some, these are not merely hyperbolic pronouncements, but practical opportunities for ending the estrangement of humans from their worlds. For, in its most visionary form, computer-driven technoculture claims to fulfill the dream of the "whole" person in the first place by healing the rupture between intellectual and manual labor. Technological utopians no longer see science and technology primarily within a discourse of social justice—namely, the freeing of labor from routine and backbreaking labor in order to free subjective time.

In the old design era, most of the architect's and engineer's time was taken up by making and remaking drawings and performing mathematical calculations. The three-dimensional graphics program and the math menu inscribed in CADD have drastically reduced the proportion of time spent on routines of drawing and calculating compared to actual design work in the activities of the engineer and architect. The consequences are already apparent. In the last five years, roughly the period of widespread introduction to CADD, hiring in civil, mechanical, and electrical engineering is confined to replacing some employees, but by no means on a one-to-one basis. Moreover, in civil engineering, owing to the broad perception that roads and bridges are seriously dilapidated, there has been a dramatic increase in public spending in construction and repair of these and other facilities. But there has been no net increase of design jobs because these employees are now more productive. Similarly, those who herald technoculture as the fulfillment of the next and perhaps final frontier of human striving call on us not to mourn the passing of the old culture—if indeed a culture not intimately linked to technology ever existed—or to celebrate a new cornucopia of leisure; they call on us, rather, to take a new ground of social existence. At last, according to some, the full development of the individual is possible, because we have finally objectified both our physical and mental capacities in a machine.

Technoculture plays with the distinction, made first by the Greeks, between work and labor. Some have argued that, in contrast to the era of mechanical reproduction, computer-mediated work eliminates most of the repetitive tasks associated with Taylorism and Fordism, that this "smart machine" can interact with human intelligence as a playmate—hence the use of chess as a test of the possibilities of artificial intelligence. Chess is the ultimate criterion in that, even if a computer cannot exceed the boundaries established by its maker, it can nevertheless provide a mirror for our intelligence. Thus, the distinction between work and play that had characterized our collective preoccupation with scarcity throughout history is sundered: for the first time, work and play are identical for occupations beyond those of artists and scientists. To borrow a term from Alfred Whitehead, technology not only "ingresses" into events, but also has become an event that leaves virtually nothing untouched. It is not a worldview in the traditional sense; instead, it is the ingredient without which contemporary culture—work, art, science, and education, indeed the entire range of inter-actions—is unthinkable. So it is futile and even deeply conservative to rail against the technological revolution. The only remaining questions are associated with how to free technology from the thrall of the organization of labor, education, and play according to the canons of industrial society. In what follows, I will make some preliminary efforts to answer these questions.

Thirty-five years ago, business and labor leaders proclaimed the "automation" revolution, an event signaled by the introduction of the transfer machine into auto-assembly plants. In 1955, during a much publicized appearance at the General Motors Cleveland engine plant where the machine had recently been installed, GM CEO Charles

E. Wilson, envisioning the fully automatic plant, is said to have asked United Auto Workers president Walter Reuther, "Who's going to pay union dues?"—to which Reuther shot back, "Who's going to buy your cars?" Now, the transfer machine was merely an electronically controlled assembly line: where thousands of workers formerly installed parts by hand (or by hand held electric tools), these operations were now performed by the machine itself. Early transfer machines simply applied to production the feedback mechanism of the thermostat, which by the mid fifties was already used in oil-, chemical-, and food-processing industries. And the machine weirdly bore a physical resemblance to human arms and hands—the prototype for which was already exhibited in that famous scene in *Modern Times*, in which Charlie is hooked up to an automatic feeding device that keeps his hands free to work while he eats lunch. The transfer machine's arms and hands are skeletonlike, bereft of flesh, but electric power multiplies the motive power once provided by muscle. In this regime of production, human labor is made ancillary to the machine and is absolutely needed only to repair it during its infrequent breakdowns.

Although the techniques of early automation were relatively elementary, at least in comparison to contemporary cybernetic practices, the fundamental purpose for its introduction into the workplace was crystal clear; later discourses about computerization, on the other hand, obscured the economic implications of computers. At the dawn of the latest technological revolution, most of those who were either its perpetrators or its objects understood what was at stake: to liberate labor from repetitive tasks and to enlarge profit margins by reducing the most expensive cost in production—manual labor. Recall the extent of unionization in the basic production industries at the end of the war: it was a historic development that established high wage and benefits levels.

On the eve of the Vietnam War, the recovering economies of Japan and Europe were already challenging, with higher technologies, U.S. industry to either modernize or close shop. However, the decade during which war industry–generated profits rolled in postponed the inevitable: when the United States turned its attention to civilian pro-

[23]

duction in the 1970s, whether too late or not, U.S. corporations seemed to lack the will to take decisive steps to save the steel, rubber, electrical machinery, and a dozen other industries; the federal government was already in the thrall of an anti-interventionist ideology that paralyzed it even as dozens of key industrial plants closed down.

Yet, dire predictions of the decline and fall of U.S. world economic hegemony proved premature. Global capitalism, still largely based in the United States, where national boundaries are no longer—or never were —sacrosanct, produced a parallel industrial regime in the global assembly line. An "American-made" automobile is likely to contain a Japanese fuel pump, a Mexican exhaust system, and Malaysian windshield wipers, or to have been assembled in Japan or Korea. The only thing "American" about many Chryslers, for example, is the name. This describes a change, the abstract name for which is the "new international division of labor."

In the 1970s U.S.-based corporations went global, sacrificing in the process huge chunks of the domestic heavy- and light-production industries. Except for truck and aircraft manufacturers, the mass-transportation equipment industry virtually disappeared as U.S. competitiveness in shipbuilding and rail cars flagged. Japanese, German, and Italian steel was routinely imported into the United States, because the domestic steel corporations resolutely refused to modernize after the 1950s, specifically rejecting computer-driven basic ingot production processes. As late as 1980, most remaining U.S. auto-parts plants were ensconced in 1920s technology, even as overseas competitors had already introduced the major computer-mediated technologies of numerical controls, robotics, and lasers—all of which were developed in U.S. laboratories but were applied only in limited ways to U.S. production industries, notably the manufacture of aircraft engines.

Curiously, in this culture that has privileged practical applications over fundamental research, the United States still retains its lead in fundamental computer-hardware components, particularly in the very crucial memory chips, partially due to massive government funding of defense-related research and development activities in the 1950s and 1960s. And U.S. entrepreneurs still dominate the computer software field, although Japan and certain European countries, particularly Italy, are beginning to catch up.

Technoculture emerges from the ruins of the old mechanical-industrial culture. From the perspective of the industrial worker, whether in the factory or in the office, the second phase of automatic production—computerization—is merely a wrinkle of disempowerment. There are no exciting new skills needed to operate the console that controls a robot; only a small fraction of people get to program the numerical controls box perched atop the lathe, and to operate lasers, for example, requires no special technical training. Similarly, much clerical labor has been transformed by the computer. The typewriter is still employed in some smaller firms, but the personal computer has all but driven it from large offices. The computer is more versatile as a word processor than its ancestor, and its computational capacities exceed those of even the most advanced mechanical accounting devices. But while the typist gives way to the word-processing operator, and the skills menu required by the new machine varies and is, perhaps, marginally richer than the older one, the operators of the most complex electronically driven and computer-controlled technologies follow routine instructions. Moreover, after two decades of the computerized office, we have already discovered its dark side: for those who perform word processing, work can be dangerous to their health. Compounding widespread diseases of muscle and bone such as tendonitis and the more egregious Carpal Tunnel Syndrome and Repetitive Movement Disorder, which cripples arms and hands, some investigators have found that the incidence of birth defects among pregnant word-processing operators has multiplied beyond the statistical average for the general population.

There is no "culture" here, just a faster line. For office workers, the computerized workplace has tended to eliminate the amenities of socialized labor. Back-office workers in large commercial banks interact with their video display terminals (VDTs) more than with other human beings. VDT operators sit in individual carrels typing nothing but disembodied numbers all day. Naturally, this numbingly boring work causes enormous turnover among VDT operators.

In the large steel, machine, and metalworking plants that have been

broadly computerized, there are fewer workers on the floor—a sometimes eerie experience for anyone of the older generation who remem-
bers the camaraderie of the age of mechanical reproduction. The
Marxist take on this—that the old factory was a prison, that the job
dulled the mind and wore out the body—is true, but it wasn't lonely;
and whether you had a diploma or not, you felt your power, because
production literally depended on what Marx called "cooperative" labor.
One need not invoke a sense of nostalgia for the preautomated assembly
line to recognize some of its ambiguous virtues.

However, the worker is no longer the subject of production in the
new technoculture because this technology is not rooted in the grand
narratives of skilled and unskilled manual work, even as its function in
the workplace (including in clerical work) is absolutely central. Even if
these narratives were purposively sundered by the division of labor, they
depended upon the great metaphor of man's conquest of nature through
the coordination of head and hand. In the regime of industrial capitalism, the "whole" producer was divided (presumably in the interest of
efficiency) and reunited by the function of management, now legitimated
as a property of capital. The socialist project consisted largely of the
promise to restore to the producer not so much the full product of his
or her labor as self-management. Now, though, the function of direction would be shared by the direct producers, who might assign the
coordinating tasks formerly ascribed to management as to any other
task in the labor process, holding neither special privilege nor powers.
At the same time, they would no longer be in the thrall of a division of
labor in which what was called "intellectual" was sharply demarcated
from what was called "manual." So even if "craft," the concept of which
is identical with the unity of design and execution, no longer described
the heart of the labor process, the formation of the detail laborer under
capitalism—which also signified relations of power and powerlessness—
would, under the new socialism, no longer be a signifier of social
power; rather, it would become merely a technical description of certain functions within the labor process.

In the past twenty-five years, computer-mediated work has been
employed, typically but not exclusively, in a manner that reproduces the

hierarchies of managerial authority. The division between intellectual and manual labor, and the degradation of manual labor that character-ized the industrializing era have been maintained, despite the integrative possibilities of the technology itself. But those who advocate it as the embodiment of the restoration of true profession or craft argue that its uses must be separated from its intrinsic character, and that it is in the interest neither of society nor of corporations to mechanically transpose the organization of work, perhaps appropriate in the industrializing era, to the new computerized workplace. On the contrary: some insist that computer-mediated work could provide unprecedented opportunities for the full development of the worker's knowledge and authority over the labor process.

On the other hand, the computer provides the basis for greatly extending the system of discipline and control inherited from nineteenth-century capitalism. Many corporations have used it to extend their panoptic worldview, which is to say, they have deployed the computer as a means of employee surveillance that far exceeds the most imperious dreams of the panopticon's inventor Jeremy Bentham, or those of any other nineteenth- or early twentieth-century capitalist.

Shoshana Zuboff, among others, insists that the computer invests communication with a new dimension by creating a richer social text, enabling people to communicate for all kinds of purposes—especially those of knowledge-sharing—to a degree that is impossible with either face-to-face interaction or print.[5] Conferencing, a technique made pos-sible by computers, permits people in widely separate places to talk to each other by creating texts. The newspaper, the magazine, and the book occupy different temporal orders from those of the ordinary workplace, in which minutes and hours rather than days, weeks, and months are the primary unit.

However, Zuboff's most important claim for the computer—that it can facilitate more communications, and particularly make for shared decision making between the higher managerial echelons and subordi-nates—remains a hypothesis. Her studies of the industrial uses of the computer demonstrate the reverse, that Bentham's panopticon still marks the worldview of the managers. The opportunities for improved communications among peers are indeed greatly enhanced by the computer, because, logically enough, peers do not take the electronic text as a command; yet she reports that one manager was mortified by the rapidity with which a suggestion communicated via computer to subordinates was received as an order. This response is, of course, conditioned by the hierarchical regime of power in the corporate workplace.

It may be concluded that the computer is janus-faced, and that it mirrors its masters, the plural indicating merely a fissure among those who developed it. Although the idea of automatic production was already part of the lore of high industrial capitalism, the self-activating loom was seen as a more practical, business-oriented application of automation than was the calculator. Latter-day Babbages, though employed by corporations that have subsumed this invention under capital's requirements, retain the utopian—anarchist, even—impulse that infuses every effort to integrate design with execution, to finally abolish the socially constructed gulf between intellectual and manual labor.

With the introduction of cybernetics into a myriad of workplaces, the outlines of a new era of paid work have already come to the surface. A recent report by the International Labour Office, a United Nations agency, estimates world unemployment at more than 800 million. Of course, joblessness may not be attributed, exclusively, to the results of technological innovation. These numbers reflect vast underemployment of most of the world, which is still locked in agriculture; the worldwide glut of goods that resulted in mass permanent layoffs in every industrialized country in the early 1990s; and the tremendous growth of the labor force, especially women, owing to rapid urbanization but also to the equally cascading living standards, especially in the United States and Latin America.

[28]   Even as we are witnessing the rapid computerization of work, the national state is undergoing vast changes. The vaunted safety-net of welfare-state capitalism is increasingly full of holes, not only in the U.S., where it was never really strong, but also in Western Europe, where once the grand compromise between labor and capital insured cradle-to-grave security in many countries. In the wake of the state's weakness, even as a series of repressive apparatuses, millions are falling through

the holes, including a significant segment of the middle class which, no less than manual workers, is experiencing the deleterious effects of computerization.

This leads to the conclusion that the current celebration of the coming of the cyborg, the possibilities of transforming labor into play, the hype about the wonderful world of the electronic superhighway and the vast horizon of the deployment of computers for music, film, and other visual arts must be tempered by the recognition that the *main* use of computers and other cybernetic technologies is to *destroy paid work.* We can rejoice at the possibility that humankind is being liberated from the oppressions of tradition, even of the newly eclipsed mechanical era. These include drudgery, boredom, repetitive mind-numbing operations, and many of the dangers of the old industrial factory. But computerization also entails the passing of a certain type of skill, that associated with the close coordination of feeling and reason, of intuition and calculation. Now the old organic self is subsumed under the cyborg self: we are wired, simulacra.

It's all very exciting, but is Philip K. Dick's dark vision more plausible than that of the new utopians? Since the great nineteenth-century industrializing era never resolved the daunting problems posed by urbanism, international migration, and the carnage left by out-of-control production, are we heirs to the return of the repressed?

Media, Technology, and the Market: The Interacting Dynamic

Herbert I. Schiller

In a slumping, perhaps declining U.S. economy, one sphere, the media-informational, retains uncontested global primacy. It has another unusual characteristic: it has been nearly totally appropriated for a sole objective —marketing. This relationship, namely, a powerful media apparatus linked tightly to salesmanship, creates a general condition that in time threatens the viability and sustainability of human existence itself. How and why this apocalyptic prospect is unfolding is the subject of the discussion that follows. The starting point is the nature of the American market in the postwar years and some of the major developments that have affected it: notably war, the militarization of science, and the emergence of a transnational business system that seeks ever more powerful ways to sell goods.

The Market in the Postwar Years

Leaving aside the drastic drop in production during the Great Depression of the 1930s, the United States was a powerful producer of consumer goods *before* World War II. During the war, the U.S. industrial machine, unmatched at the time, concentrated on the production of war goods: tanks, munitions, aircraft, military supplies, and so on. But the Depression remained a vivid memory, and few would have predicted the continued expansion of industry that followed the hostilities. In fact, many were anxious the economy might recede into the prewar economic crisis.

Withal, the deferred domestic demand and the forced savings accumulated during the war years—for example, war bonds—pushed output in America to unprecedented levels. To this was added the orders for goods to replenish Europe, stripped bare during the Nazi occupa-

[31]

tion. Accordingly, the industrial plant that had been built up during the war and later converted to civilian production, supplemented with massive capital investment and expansion of industrial capacity in the early postwar years, provided a powerful productive base. From it flowed a huge volume of consumer goods in the late 1940s, '50s, and '60s.

The ability of Americans to buy these goods kept up with their output at least until the early 1970s. U.S. personal consumption expenditures (in constant 1982 dollars), fueled by the hot wars—Korea, Vietnam—and the Cold War in general, rose from $503 billion in 1940 to $2,682 billion in 1990, a fivefold increase over the fifty-year interval; the population merely doubled in size in the same period. By decade, the percentage increase in personal consumption expenditures was as follows: between 1940 and 1950, 46 percent; between 1950 and 1960, 37 percent; between 1960 and 1970, 49 percent; between 1970 and 1980, 34 percent; and between 1980 and 1990, 34 percent.[1]

The income to sustain these rapidly rising expenditures was heavily supplemented with the growth of the credit industry, which discovered all kinds of imaginative ways to induce the acceptance of staggering personal debt burdens. Completing the infrastructure of the consumer society in the making was the parallel expansion of the advertising industry and advertising expenditures over the period.

The advertising industry is the chosen instrument of the unplanned yet heavily concentrated American economy to clear the inventory off of corporate shelves. As a special bonus it also supplies strong and pervasive doses of systemic ideology. This is more flatteringly expressed by advertising authority Robert Coen:

> Unlike the European societies, the U.S. had no established rigid marketing system, guilds or hierarchies, and U.S. entrepreneurs were not constrained in developing their enterprises. They discovered that they would prosper if they spread the word about their goods and services. The advertising industry, as known today, has mainly evolved from the American experience.[2]

It is precisely because American entrepreneurs historically have had fewer constraints that this model of economic development, in its purest capitalist form, has occurred in the United States. And in this model, advertising occupies a central position. It is to be expected and, in fact, is confirmed that "in the case of advertising, most of the advanced developments have occurred first in the United States."[3] That this continues to be the case will be made evident further along.

For advertising to fulfill its systemically crucial role — getting the national output of goods and services into the hands and homes of buyers, and reaffirming daily, if not hourly, that consumption is the definition of democracy — it must have full access to the nation's message-making and message-transmitting apparatus. Over time this means the transformation of the press, radio, television, cable, and every such subsequent technology into instrumentations of marketing.

This is done with single-minded devotion. It has succeeded so well that the nation's image- and message-making machinery has been almost fully directed to salesmanship. The press is dependent on advertising for approximately three-quarters of its income. Commercial radio and television are totally reliant on advertising revenues, and the public channels, too, are increasingly dependent on this source of support. As each new media technology has appeared, it has drained off some of this marketing income. Broadcast and cable television are now the primary, though by no means exclusive, vehicles for the sales message.

All of this is well known, amply documented, and analyzed in trade and academic literature. What the record reveals is an almost total takeover of the domestic informational system for the purpose of selling goods and services. What is not so familiar, however, is the extent to which this marketing function has influenced the form and shaped the content of the new information technologies to suit its purposes. This can better be appreciated by reviewing the history of the new information technologies that have arisen since World War II.

[33]

War and Business: Recent Stimulants to Technological Development

The production of imagery and messages relies on institutional structures and relationships, as well as on instrumentation. What is produced is decided upon by a complex interaction of processes, techniques, and social relationships, all of which are in constant flux with changing weights for each component. The U.S. message and image machine, in response to two separate but related forces—military and corporate—has developed very special features over the last half-century.

Henry Luce's "American Century" was a corporate vision, revealed to and proclaimed by Luce in 1941. It was a design thoroughly dependent on a greatly expanded international network of communication. It also foretold the placement of armed forces around the globe to defend the emerging system of American economic power and ideological persuasion.[4]

When the Cold War ended (whenever that date is authoritatively established), the United States maintained 375 foreign military installations worldwide, staffed by more than a half-million service people.

It also had a powerful scientific and technological establishment that had been created in World War II and expanded in the postwar years. This decisive element of national power was the offspring of the astronomical Cold War budget that for more than forty years channeled massive funds into scientific research and development. These sums poured into federal, corporate, and academic installations. A science reporter put it this way:

> By some measures, the cold war was the best thing that ever happened to research. The explosion of money, talent and tools far exceeded anything in previous eras. Over the decades [an] army of government, academic and industry experts made the breakthroughs that gave the West its dazzling military edge. Since 1955, the Government has spent more than $1 trillion on research and development of nuclear arms and other weaponry.[5]

From this almost inconceivable outlay came, among other products, laser weapons, spy satellites, precision armaments, weather satellites, computer chips—in fact, the reporter asserts, "the whole industries of aerospace, communications and electronics."

Out of this proliferation of scientific and technological projects of military and corporate parentage has also come what is reassuringly called the "information society," in which we are now living. Yet the main beneficiaries of the new capabilities in information production, transmission, and dissemination are, not unexpectedly, those who were the main initiating agents of the Cold War era—the transnational corporations, the intelligence, military, and policing agencies. Especially well rewarded have been the big businesses with worldwide operations. With new facilities they have the means to manage their global activities, move capital, shift production locales, and, on the basis of these new capabilities, weaken organized labor. At the same time, the Pentagon and the intelligence agencies have built intercontinental satellite networks for monitoring the flow of messages of friends and foes alike, and for mapping the world for possible future interventions—Iraq, Libya, North Korea, Cuba, inner-city Detroit, and so on.

While this new technological capability might have been expected to strengthen and perpetuate American domination of the international realm—and for a brief time it did—other developments intervened, notably the war in Vietnam, to erode U.S. power. In recent years, the growth of rival industrial systems in Germany and Japan has further diminished American influence around the world. *But not in all sectors—* especially not in the increasingly central sphere of communications.

The new information technologies enhanced the vigor and production quality of the American cultural industries, conferring on them a marked technical edge. But this was no random transfer of technical expertise: a familiar and pervasive force, the marketing imperative, initiated and guided the adaptation of many of the new technological processes into the cultural industries. The process responded closely to the growing pressures on the overall American economy.

In the first postwar decades there was scant need for American marketers to take into account either the domestic or the global buyers' tastes and preferences. American goods had few competitors. The early postwar boom years enabled U.S. products to be marketed with effective but (by current standards) unexceptional advertising. American media products were pumped into the world market with few constraints.

These "good times" for American industry began to falter in the late 1960s. Though at first a seemingly temporary irritant, the pressure

[35]

on the economy of renewed foreign industrial competition, along with the exhaustion of the postwar replenishment boom, continued to build. In the late 1980s and early '90s it has become relentless. The slowdown of industrial production, so clear domestically, is occurring across most of the industrialized world as well.

New Means for Bolstering the Marketing Message

The marketing imperative has taken on a new urgency. Beginning in the 1970s, the search for consumers became increasingly feverish. Messages addressed to potential consumers have multiplied and intensified. As the overall U.S. power position deteriorates, its one sphere of unchallenged mastery, media-driven pop culture, receives mounting dosages of adrenalin from technologies and processes developed originally to serve the empire. Now they are aimed at producing consumers. Popular culture is, in fact, fusing with merchandising, and adapting and using the instrumentation and techniques that appeared earlier in the research labs of the cold and corporate warriors.

Indeed, an entirely new economic sphere, the special-effects industry, has emerged, difficult as it is to place into the standard industrial classification. What, for example, is the product of George Lucas's commercial special-effects company, a Hollywood shop aptly called Industrial Light and Magic? Yet it is not some hidden or mysterious sphere. As the screen credits for any typical high-budget film unfold, a battalion of names appears in one or another highly specialized fields, generically now considered "special effects."

The use of special effects long precedes the current period. But although they were certainly evident in the early development of film, the elevation of special effects into a primary constituent of film is much more recent. It can probably (and arbitrarily) be assigned to their use in *Star Wars*, produced in 1977, although a case could also be made for Stanley Kubrick's *2001: A Space Odyssey*, which, when it was made in 1968, achieved a startling pseudo-environment.

The actual date that marks the full emergence and use of special effects in the cultural industries is uncertain at best. It is sufficient to note that sometime in the post–World War II period, centered around the late 1960s and early '70s, the provision of special effects became a powerful and observable component of popular cultural forms and products. Though no one factor explains this development, and several have believability, it is the contention here that one contributing stimulus stands out: *the intensifying effort to maintain and extend personal consumption and to develop it abroad.*

In a word, the big consumer-goods producers, a.k.a. "the national sponsors," and their advertising agencies have been pressing unremittingly to target consumers with ever-more powerful sales messages.

It is probably true that this many-pronged effort would have been undertaken routinely, in the day in–day out operations of corporate business as it continuously jostled for market share. But the additional pressure, and the time in which it has become manifest, coincide with the onset of fierce global industrial competition and creeping domestic economic stagnation.

Conveniently, there were available, thanks to the trillion-dollar research and development expenditures, a constellation of new techniques and technologies—some applied to other uses—that could satisfy the most extravagant expectations of corporate marketers and salespeople. These technical capabilities for use in visual, optical, photographic, and audio forms offered improved ways to intensify excitement, organize audience attention, and capture and hold interest. What better way to describe the objectives of advertisers on behalf of their sponsors, or of record producers hoping for platinum disc sales, or movie producers seeking box office blockbusters?

The special-effects industry is an extraordinarily acquisitive and derivative industry that actively searches out advanced technologies from elsewhere—the sciences, medicine, computer graphics, personal computer developments—and puts them to astonishing use. In a related way, the marketing effort was applied to stimulating personal acquisition of a growing number of electronics products that became available at about the same time. This served, conveniently, to make existing record collections and record players, for example, obsolete. The introduction of personal computers on a mass scale, the phenomenal popular embrace of VCRs, and the promotion of compact discs opened up new consumer markets and launched new entertainment products. These new electronic goods also enabled individuals to draw spectacular techniques *into their own living rooms.* Looming now is High-Definition TV (HDTV) in the parlance. HDTV, if and when it is introduced, promises to make America's existing stock of television sets—hundreds of millions—instantly out of date. Planned (or not) obsolescence with a vengeance.

And so it has developed. In recent years, TV commercials, movies, TV programs, and recordings have called increasingly upon special effects to rivet an audience's attention. Special-effects sound and imagery short-circuit the brain and hit the gut. Content recedes as technique flourishes and reflection disappears. This may be the defining feature of what is called postmodernity.

Philip Hayward writes about these developments in music video and MTV, where he sees "the impact of special effects as stimulating and retaining audiences" and bypassing "concern with textual profundity in favor of technological 'magic'and rewarding highly competitive forces."[6] More pointedly still, another writer sees the ultimate effect of special-effects techniques in pop-cultural forms as the triumph of commercialism:

> The wedding of pop music and video has shifted the balance of power between a song and the image of its performer. When once that image augmented the music, nowadays the music serves the image, which in turn, often serves a fast-food or cola advertising campaign and ultimately a whole line of advertising. Pop stars have always endorsed products, but it has been only in the age of music video that the star image and the product have become indissoluble ... pop music is no longer a world unto itself but an adjunct of television, whose stream of commercial messages projects a culture in which everything is for sale. As the image has taken precedence over the song, the songs have become fragmentary and charged with electronic beats.[7]

The writer, Stephen Holden, concludes from this that the "wondrous technology" has "tricked us." Yet is this galloping phenomenon merely a matter of smoke and mirrors? Are special effects just surface phenomena that willy-nilly contribute to confusion and mystification? More convincing in explaining why this has been happening, I believe, is Martha Rosler's on-the-mark observation that "the confusion of style with substance is fostered by any medium that allows advertising to be integrated into its fabric and format."[8] Is not what Rosler calls attention to exactly what has been occurring across all media? Advertising, with the closest cooperation of the new technologies, has been integrated increasingly into film, television recording, and news itself.

Admittedly, this trend preceded the new technologies. In film, for example, the widespread use of the "built-in plug," examined so deftly by Mark Crispin Miller,[9] allows actors to clutch a can of Coke or a Bud or cleaning fluid, seemingly as part of the natural setting. More generally, many movies, in addition to routinely and endlessly promoting consumption, continue to offer the bromide of rags-to-riches sexual liaisons with humanistic billionaires transformed into social workers who disinterestedly assist working-class women — *Batman, Pretty Woman, Working Girl*, and so on. Two systemic objectives are realized with this formula. Along with the consumerism — almost as a bonus — comes a happy escape from economic anxiety and deprivation. The long-running American dream show is revived and re-energized.

In another way, television from its inception has had its drama and sitcom programs written around the commercial break. The climaxes are scheduled carefully either just before or just after the sales message.

Now the new technologies enable imagery to take a giant step beyond this mode of commercialism. In recent years the actual form of a presentation is structured for a commercial objective — to hold the viewer/listener captive to the product message. MTV is the purest example of this development: "'Atmosphere advertising' is the term for ads driven by music and visuals, with no hard sell."[10] MTV is nothing more than an advertisement for musical groups and record labels. Stuart Ewen points out that "part of the magic formula embedded in MTV is the fact that viewers [are] looking at an advertisement without really noticing it."[11]

The impact of MTV is not limited to advertising, important as it is in this domain. It has served as a successful model for the rest of the media as well, including print. *USA Today*, for example, offers its news in one or two small paragraphs, print capsules surrounded with color copy. The brief soundbites of television news reporting get shorter from year to year. One MTV "news program" for young viewers—MTV's basic audience—boasts three-second interviews with political figures.[12]

As the information gets compacted, so does the general programming. Time-compression machines, introduced in the early 1980s, are used routinely to "make room for more commercials by accelerating the speed of movies and old television programs." Their use is especially attractive to broadcasters because "the additional minutes of advertising time come without increased costs for additional programming." This technique "can cut 8 percent of a two-hour movie by speeding it up [and] gain almost ten minutes without cutting any scenes."[13] What this practice does to the integrity of the original film or TV program is not a consideration for those benefiting from the sale of more commercial time.

The application of special effects to film, TV, and recording in the effort to secure viewer/listener attention for a commercial message results inevitably in the evacuation of meaning. One film reviewer writes, "One of the most numbing things about many movies today is how wildly out of scale they seem to go: the way visual and technical virtuosity is juxtaposed with silly, vacuous stories."[14] The same reviewer reflects further on the usage of special effects in film: "There's something dispiriting about the wide disparity in quality between words and images in current American film-making." Writing about a recent movie, *The Lawnmower Man*, it is noted that "every scene in this cautionary tale about science running amok has spectacular views, unusual camera angles and moves, or dazzling outré computer effects. And every scene, story-wise, gets mushier and more outlandish or perfunctory."[15]

This could be said as well about most of the razzle-dazzle film productions that run up huge budgets. Indeed, special effects techniques are more and more often applied "wall to wall" in order to sustain the desired sensation throughout the entire film, TV program, or recording—notable examples being *Blade Runner*, *Terminator 2*, *Roger*

Rabbit, and *Jurassic Park*. Yet this inverse relationship between technique and content arises directly from the commercial imperative. If the primary aim of the sponsor is to capture the attention of the audience with an emotional jolt, why distract or depress them with a serious story line or lyric?

Export of the American Model

Media-cultural developments in the United States preview their adoption abroad. This occurs in the "normal" course of transacting global business. The very expression "world market" now means that globally operating corporations produce and sell their wares in scores of countries. Most of the market is shared by a handful of giant companies in specific spheres—electronics, recordings, cereals, home products, automobiles, pharmaceuticals, and so on. Following this pattern, it is possible to detect in recent years a complex unfolding of media, technological, and market relationships across a good part of the world market. It is clearest in Western Europe, Japan, Australia, South Korea—in fact, wherever a national economy has reached a certain level of industrial production.

Each national locale has its own specificities in this evolving pattern, but the common course of development, at least approximately, can be sketched. It runs something like this: Advanced communications technologies—including computers, satellite communication, broadcast, and cable television—have become available. The transnational corporate sector, with its supporters inside each national configuration, presses for and succeeds in achieving what is called "deregulation" or "liberalization" or "privatization." One way or another, these variants remove national oversight and open the door to commercial use of broadcasting and telecommunications. This means financing TV and cable increasingly by advertising, and the private (corporate) use of telephone lines and computer networks for heavy-volume (business) data flows. With these arrangements in place, the framework is established for the creation and utilization of cultural space in the American mode. And this is the way it has gone! One media-cultural sphere after another has been seized for corporate marketing goals, sometimes by way of media product imports from the United States, sometimes by imitating American pop-cultural production styles, often a combination of the two. A report on European TV developments is illustrative: the Hollywood-focused *Los Angeles Times* described a new European TV soap opera, "Riviera," as having a "decidedly American feel to it." No wonder! The executive producer and creative director of the show are American. "U.S. consultants were hired and brought to Paris for sets, music, lighting and engineering ... All taping is done first in English before the show is dubbed into French and other European languages. Yet most of the actors, directors and studio technicians are Europeans. [The show was] conceived in a Paris advertising agency." The French producers acknowledge that "the Americans know how to capture and hold an audience. This business can benefit from tricks invented in Hollywood. Young French writers [for example] are taught how to structure plots in the American way, which means inserting mini-climaxes before commercial breaks."[16]

Not only are European TV and film yielding to—perhaps welcoming—the marketing call. The total packaged commercial environment also is becoming familiar: witness Disney's opening of a huge theme park outside Paris in April 1992. *Business Week*, only partly joking, described the event: "And you thought Europe's big happening of 1992 was the creation of a borderless market. The event that really counts is coming April 12. That's when Europe will join the U.S. and Japan as a truly advanced consumer society: It's opening day for Euro-Disneyland." Hyping the opening, a $6 million newspaper, funded partly by corporate partners including Mattel, Coca-Cola, and American Express appeared in seven major cities. *Business Week* asks, "How can Europe resist? Chances are it won't. Cultural imperialism or not, millions of Europeans will soon be careening down Big Thunder Mountain, celebrating the Old World's coming of age."[17] Most of these installations *are* notably proficient in shepherding the happy throngs who pay for the privilege of being exposed to commercial messages through artfully integrated and carefully constructed space.

In 1988, MTV European satellite service was introduced. This network brings together the communication satellite, cable TV, and advanced visual and sonic special effects to transmit the marketing message to an audience in the most effective way. One of the first shows it carried claimed this distinction: "We see *Ultratech* as a very global idea because there is no language, there is no plot, there are no characters....It's just sound and light, TV taken down to its essence."[18] In other words, the perfect sales tool! With deregulation achieved, cable and satellite TV installed, American special effects techniques imported or copied, the Europeans, and many Asians as well, can look forward to world-class—that is, American—levels of commercialism. The international advertising industry is already drooling: "On a per-household basis, four times as much is spent in the United States on television advertising as in Europe. The potential for growth in television advertising revenues in Europe is [seen as] enormous ... the industry still looks like a pot of gold to many."[19]

But is it only gold that is at the end of the advertising rainbow? Not yet fully visible but certainly accompanying the short-term profits is the escalating waste of limited natural and human resources. As car sales rise, gas consumption soars, highways encircle the earth, and Coke bottles and disposable wrappers lie alongside the road, cultural pollution, too, gushes forth from the global cultural industries.

Autonomous and Neutral Technology?

Calling attention to these developments, it should be emphasized, is not intended to sound still another antitechnology warning. "Technophobia," as Neil Postman put it, is not the issue.[20] What the evidence here demonstrates is the strong, if not determining, influence of the social purpose that initially fostered the development of new technologies. The social uses to which this technology is put, more times than not, follow their originating purposes. When military or commercial advantage are the motivating forces, it is to be expected that the laboratories will produce findings conducive to these objectives. If other motivations could be advanced—for instance, the general welfare—different technologies might be forthcoming. The customary argument that commerce and profit-seeking go hand in hand with social benefit is yet to be demonstrated—after hundreds of years of experience.

Robert Kubey and Mihaly Csikszentmihalyi explain television's alleged *inherent* technological imperative this way:

> This is a good place to debunk the much-repeated idea that television is a medium best suited to transmitting emotions, and that it either "cannot" or is not "good" at transmitting ideas. The answer to why we see what we see on television lies in a combination of how audiences have come to conceive of the medium, what audiences want to watch (or have grown accustomed to watching), and what the people who control and sponsor television believe needs to be created and broadcast in order to maximize profit.[21]

Each one of their criteria clearly is a social construct. They are not the outcomes of an autonomous technological determinant. The TV that is transmitted, the movies that are made, the music that is recorded and played, and the images that have been captured and displayed increasingly are shaped to serve the marketing imperative. This driving force, most pronounced in the United States, is now being experienced in much of the rest of the world.

To recapitulate: The strongest sector in an otherwise declining U.S. economy is the cultural industry. Its exported product constitutes one of the few favorable components in the country's foreign trade. Additionally, the styles of the industry provide the model that is copied everywhere. Seemingly, this great success, and the rich returns that flow from it, should be a source of satisfaction and security; yet a very different reaction is warranted, and therein lies a paradox. A buoyant cultural industry is responsible for a looming social disaster.

The media-entertainment industries flourish; their revenues rise from year to year, and their value to the consumption goods industries continually increases. But the virus of mindless consumerism accompanies—in fact, is embedded in—the cultural product that now reaches nearly every corner of the world. The waves of merchandising exhortation carried by, and built into, the popular media crash against the earth's finite resources. Whether it be the disposition of atomic wastes, the widening holes in the ozone layer, or the desperate search for waste

[44]

disposal sites, underlying each crisis is the misuse of resources. In the fall of 1991 a group of world-renowned intellectuals, attending a symposium titled "Approaching the Year 2000," indicted the waste and despoliation of the earth's resources, noting a grotesque distortion in global resource allocation whereby "20% of the world's population consumes 80% of its wealth and is responsible for 75% of its pollution." The symposium found the source of this shocking phenomenon to be "the cultural pollution and loss of tradition which have led to global rootlessness, leaving humans, through the intensity of mass-marketing, vulnerable to the pressures of economic and political totalitarianism and habits of mass-consumption and waste which imperil the earth."[22] The main miscreant in this deepening global crisis is the model of acquisitive behavior and consumerist attitude constructed and circulated worldwide by the powerful and deadly combination of media, technology, and the market.

I would like to acknowledge the research assistance of Judith Gregory.

From Virtual Cyborgs to Biological Time Bombs: Technocriticism and the Material Body

Kathleen Woodward

> Technology discloses man's mode of dealing with nature, the process of production by which he sustains his life, and thereby also lays bare the mode of formation of his social relations and of the mental conceptions that flow from them.
>
> —Karl Marx (1887)

> Ageism is usually regarded as being something that affects the lives of older people. Like ageing, however, it affects every individual from birth onwards—at every stage putting limits and constraints on experience, expectations, relationships and opportunities. Its divisions are as arbitrary as those of race, gender, class and religion.
>
> —Catherine Itzin (1986)

The dominant narrative of the historical development of technology that emerged from the technocriticism of the 1960s, '70s, and early '80s was grounded in the cluster of technologies associated with the so-called "revolution" in communications—the electronic media, the computerization of the workplace and other spaces, and the cybernetics of war. The postindustrial prophets, as they were called by William Kuhns in 1971—Marshall McLuhan, Norbert Wiener, Buckminster Fuller, and Harold Adams Innis, among them—were all noted for their fervent utopian and dystopian visions of technological change (today we might also note that these prophets were all male).[1] Information was singled out as the primary commodity of what was variously called the

[47]

"technological society," the "postindustrial society," or simply, the "information society." Since the 1970s these terms have waned considerably in their effectiveness. Nevertheless, the power of this story persists.

Witness in the popular press, for example, *Time* magazine's decision to make CNN's Ted Turner the Man of the Year in 1991. Turner himself was, of course, merely a surrogate for the corporate institution that was being lauded as an unprecedented agent of social and cultural change—the network of global communications satellites. The tradition of American individualism requires a human hero, hence the human "head" of CNN was featured on *Time*'s cover, not a graphic representing the technology itself. Or consider an April 1992 *Newsweek* cover story trumpeting "Computers to Go" as the "Next Electronic Revolution" in its headlines. It portrayed two young male tech-nerds as representative heroes—a ludicrously debased notion of revolution if ever there were one.

In recent years we have also seen the "return" of communication technologies in the writing of what I refer to as the second generation of academic technocritics, most prevalently in cultural studies. They include, for example, the prominent French intellectuals Paul Virilio and Jean Baudrillard; the former has trenchantly critiqued the cybernetic techniques of visualization that are used in contemporary warfare (specifically, the Gulf War) while the latter has extolled the power of the mass media to produce simulacra—as well as a powerful and passively aggressive silent majority. In the United States a plethora of books have recently appeared that focus on what we might broadly call cyberculture, including Avital Ronell's *The Telephone Book: Technology—Schizophrenia—Electric Speech* (1989), Mark Poster's *The Mode of Information: Poststructuralism and Social Context* (1990), Constance Penley and Andrew Ross's edited volume *Technoculture* (1991), George Landow's *Hypertext: The Convergence of Contemporary Critical Theory and Technology* (1992), Scott Bukatman's *Terminal Identity: The Virtual Subject in Postmodern Science Fiction* (1993), and Andrew Ross's *Strange Weather: Culture, Science, and Technology in the Age of Limits* (1991). Echoing the technocritics of the 1970s, Ross designates the information revolution as "the chief capitalist revolution of our times," observing in a post-

McLuhan tone that "the cybernetic countercultures of the nineties are already being formed around the *folklore of technology*—mythical feats of survivalism and resistance in a data-rich world of virtual environments and posthuman bodies."[2]

There was, however, another story that was told in the formative years of technocriticism in the United States, a narrative rooted not in communications technology but rather in biotechnology. Why did the major technological character in this story not receive equal attention? What versions of this narrative are being told today? What role does gender play in the technocriticism grounded in biotechnology? How do these two narratives of technological change—the dominant narrative based on communications and cybernetics, the recessive narrative based on biotechnology—figure the human body? These questions are central to what follows.

But first I want to briefly note several of the overlapping assumptions that underlie my thinking about technology: one, that technology is not "neutral"; two, that it is absurd to think of technology as "out of control" or "autonomous" because it is in fact thoroughly embedded in social processes; three, that it is preferable to remain as concrete as possible when we think about technology, referring to particular technologies in specific contexts (solar energy for a home in a suburban setting, for instance, or laser brain surgery for a child in a high-tech medical complex) rather than to Technology as a monolithic demonic or liberating historical force; and four, that it is important to recognize when the term "technology" is being used primarily metaphorically to refer to something other than itself (as when, for example, Michel Foucault uses "technology" to refer to the *discursive workings* of a particular cultural formation).[3] We must be careful, in other words, to tease out the ideological implications of technocritical thought and rhetoric. Specifically, I will argue here that, in general, discourse about technology in Western culture is fundamentally bound up in a rhetoric about age, one that has negative—if for the most part subliminal—consequences for all of us.

One way of reading the narrative of technological development based on communications that has been elaborated over these two generations of technocritics is as a story about the human body. It is a story of an inevitable evolutionary process mapped on the anatomy and physiology of the body, one that has a surprising ending. It goes like this: Over hundreds of thousand of years the body, with the aid of various tools and technologies, has multiplied its strength and increased its capacities to extend itself in space and over time. According to this logic, the process culminates in the very immateriality of the body itself. In this view technology serves fundamentally as a prosthesis of the human body, one that ultimately displaces the material body, transmitting instead its image around the globe and preserving that image over time. (Today that image of the body is itself capable of being altered with the help of any number of software programs.) What began slowly has proceeded at an ever-accelerating pace. If we understand the agricultural revolution in terms of the extension of the arm by the tool, and the industrial revolution in terms of the augmentation of the power and dexterity of the human body as a whole with complex machines, then it follows that the postindustrial revolution is defined in terms of the extension of our minds by information technologies. In McLuhan's ecstatic phrase, the media serve to extend "our central nervous system itself in a global embrace, abolishing both space and time as far as our planet is concerned."[4]

It is paradoxical—seductively so—that while the new communications and cybernetic networks permit increased visual access to far-flung parts of the world as well as to the inner recesses of the human body, they are based on technologies that are "unseen." There is a beguiling, almost mesmerizing relationship between the progressive vanishing of the body, as it were, and the hypervisuality of both the postmodern society of the spectacle (Virilio) and the psychic world of cyberspace. As the philosopher Hans Jonas has insisted in his work on technology, electricity is "disembodied, immaterial, unseen"; electronics "creates a range of objects imitating nothing" (the satellite is a perfect example of this).[5] How can we account for this fascination with the simulacra of postmodernism, with copies without originals, with the paradoxical insubstantiality and invincibility of the body in the space of virtual reality?

In 1930 Freud, in a passage rare for him, suggested a compelling and time-honored answer to this question. With "every tool man is perfecting his own organs, whether motor or sensory, or is removing the limits to their functioning," he wrote in *Civilization and Its Discontents*. For Freud, technological "progress" represents the fulfillment of a deeply-held psychic wish—that "Man ... become a kind of prosthetic God."[6] From a psychological, if not psychoanalytic perspective, then, the possibility of an invulnerable and thus immortal body is our greatest technological illusion—that is to say, *delusion*. This desire for the impregnability of the body is expressed unmistakably in the scenarios composed for countless video games. It is played out, over and over, in virtual reality—and in the stories that are written about it.

A key text that illustrates this with consummate if not bizarre clarity is the cult science-fiction film *The Lawnmower Man*, directed by Brett Leonard and released in 1992. *The Lawnmower Man* is a contemporary version of Daniel Keyes's 1966 novel *Flowers for Algernon* (a narrative perhaps better known in its film incarnation as the 1968 *Charly*, starring Cliff Robertson). It betrays, however, none of the moving poignancy of the conclusion of *Flowers for Algernon*. In *The Lawnmower Man* we find instead a demonic 1990s version of McLuhan's utopian prediction of a global village united through communications technology.

In *The Lawnmower Man* a good-looking youth named Joe (Jeff Fahey) becomes part of an experiment to stimulate the growth of learning, an experiment designed by an ambitious but misguided scientist named Dr. Angelo (Pierce Brosnan). Joe is what we used to call retarded. Amazingly, his capacity for cognitive learning increases exponentially during the course of the experiment. He is injected with drugs. More significantly, he is hooked up to a high-tech virtual-reality trapeze in the scientist's government laboratory, aptly if unimaginatively named Virtual Space Industry. The once sweet and naive Joe rapidly surpasses his teacher Dr. Angelo in mental ability. As he does, his appetite for knowledge and power—the two are inextricably linked in *The Lawnmower Man*—grows insatiable. (Interestingly enough, his appetite for sex, which is stimulated at the beginning of the experiment, appears to dwindle as he comes to desire instead domination over the entire

[51]

globe.) Joe takes over the administration of the experiment himself, displacing his teacher. Enabled by his enhanced mental abilities to *will* the destruction of others, his taste for bodily violence reaches horrific proportions. He uses his hastily acquired and formidable mental skills to ruthlessly murder people. He pulverizes their bodies into pixels. In a hilariously gruesome scene he turns a "virtual" power lawn mower against a neighbor he has learned to hate, cutting his body to shreds.

Fairly early on in the narrative Dr. Angelo says, "Sometimes I think I've discovered a new planet, one that I'm inventing." For Joe this "planet" is the psychic "universe" of virtual reality. With his hypermental ability he is ultimately capable of interfacing his mind with the global telecommunications network, expanding the definition of the universe of virtual reality to hyperbolic proportions. Joe will allow no one to interfere with his access to this universe—which in *The Lawnmower Man* literally means control of the entire planet. The film concludes with the image of Joe, dressed in his skin-tight, silver-and-grey body suit, as he takes on the demonic persona of his imaginary two-dimensional body in virtual reality—and leaves his organic body behind. He merges triumphantly with the interconnected communications system that circles the globe, never to emerge again in human form.[7] Corresponding to his contempt for the bodily integrity and suffering of others is his overweening desire to "peel back" the layer of his own organic body, to rid himself of what has become its intolerable limits, the organic body itself. What Freud referred to as a "prosthetic God" is figured in *The Lawnmower Man* in terms of overpowering evil. We are given to understand that the mind of Joe is now omnipresent, permutated around the globe, that he will exert dictatorial control through telecommunications technology.

[52]   I've alluded to Freud again because his work inevitably returns us to the reality principle, which in the context of this essay is represented by the very *materiality of the human body*—the technological possibilities for altering the body itself. As I suggested earlier, in the late 1960s and throughout the '70s many did insist—and I agree with them—that the most significant and far-reaching technological revolution on the horizon is predicated on biotechnology, not communications technology.

This revolution entails not the mere extension of the body and its images, but more fundamentally, the saturation, replication, alteration, and creation of the organic processes of the body—if not the very body itself—by technoscience. Genetic engineering and reproductive technologies are two prime examples. (It's important to note here that the way these technologies intervene in the biological processes of the body is radically different from the way discursive formations, as Foucault has so persuasively demonstrated, construct and discipline the body.)

Why was this important story not reproduced in the academic marketplace of technocriticism in the '70s?[8] First, it was almost exclusively a depressingly dystopian discourse; its shrill predictions of the disastrous breakdown of the "natural" law and order of the body were not counterbalanced by utopian claims that could have led to productive and tempered debate. It is notable that many of the books that dealt with the biotechnological revolution carried what were essentially warning titles—for example, *The Biological Time Bomb* (1968), *The Biocrats* (1970), *Genetic Fix: The Next Technological Revolution* (1973), *Pre-Meditated Man: Bioethics and the Control of Future Human Life* (1975), *Prolongevity* (1976), *No More Dying: The Conquest of Aging and the Extension of Human Life* (1976), and *Bio-Babel: Can We Survive the New Biology?* (1978).

Second, its concerns intersected with those of medicine, which in the U.S. has generally been thought to be a more "private" or "personal" matter, not one that requires "public" debate or policy. (This is, of course, a faulty assumption, as today's debates on national health care demonstrate; moreover, it should be abundantly clear to us that no public policy at all is in fact a policy.) And it is indeed true that in any of a number of these books one will find chapters devoted to issues such as patient consent and the right to die, issues that may have nothing to do with biotechnology explicitly and are generally categorized as questions of medical ethics.

Third, in contrast to the study of communications technology (the media and cybernetics), debate about biotechnology did not find an academic niche in the American university of the 1970s. During this decade many undergraduate interdisciplinary programs were established in the U.S. to develop curricula that would address the study of technology from the perspective of the humanities; the ideal was often cast as a partnership between faculty from engineering (definitely not the sciences) and faculty from the humanities and social sciences (history, philosophy, and literature were the disciplines most widely represented). Known as science, technology, and society studies (STS), these programs devoted virtually no attention to issues of biotechnology. Thus, for example, in the program called Cultural and Technological Studies, in which I taught at the University of Wisconsin, Milwaukee, in the late '70s, only two courses out of some fifty dealt with biotechnological issues (and these courses addressed problems of population control, medical ethics, and ecology as well as genetic engineering).[9]

And fourth, I suspect that biotechnology did not receive the sustained scholarly attention it deserves because many of its most spectacular results, as well as its more mundane concerns, were associated with women—with motherhood, children, and care-giving. I am thinking, for example, of test-tube babies and birth-control devices, of surrogate motherhood and the biotechnological age of "viability" of "fetuses," of the right to die and life-support equipment.

It is altogether fitting, then, that it is feminists who have taken up issues raised by biotechnology, focusing on gender and the technology of *reproduction* (not *production*)—that is, on the politics of biological reproduction. Ranging from Shulamith Firestone to artist Valie Export, many have argued that these new technologies give us the possibility of seizing control of reproduction. As Firestone approvingly put it in her book *The Dialectic of Sex* (1971), women can be freed from the biology that has tyrannized them; the bearing of children can be "taken over by technology."[10] Yet many women, not to say men, find this prospect threatening. Here, for example, is Richard Restak on biotechnology in 1975: "the list of potential horrors—in vitro fertilization, anonymous sperm banking, behavior-control technology in the prisons and ghettos, experiments on unwitting human subjects—is lengthening."[11] That Restak would conflate in-vitro fertilization and anonymous sperm banking with the horrors of experiments on unwitting subjects is, of course, telling.

In the United States it is, however, feminist biologist Donna Haraway's wide-ranging and brilliant "A Cyborg Manifesto," a piece that appeared a decade ago, that has been the most seminal in articulating the importance of biotechnology in contemporary cultural criticism. Haraway achieves this in great part by smuggling, as it were, the subject of biotechnology into an essay that deals with communications technology, ever understood as the hallmark of a postmodern world. In mapping what she calls the current "movement from an organic, industrial society to a polymorphous, information system," Haraway points to the shift from organic sex-role specialization to optional genetic strategies as one of its major features.[12] Haraway's essay is theoretical science fiction; she envisions a convergence of biotechnology and communications technology in the figure of the cyborg. A hybrid of organism and machine, the cyborg represents for Haraway the possibility of a world that is postgender—for her, arguably its most important aspect. Like much literary science fiction at its best, Haraway's essay is a critique of the present, one that is markedly clear in its acknowledgment of the social and historical constitution of gender and race.

But what of that other ubiquitous and virtually unanalyzed marker of the body—age? I am concerned here, in other words, primarily with the aging body, not with the sexual geography of the body (which is in any case being altered by surgery—I am referring, of course, to sex-change operations). Much recent research on the body in what is broadly referred to as cultural studies has been concerned with the ideology of difference, with understanding how differences are produced by discursive formations, social practices, and material conditions. And yet in cultural studies the only bodily difference that has received little scholarly attention is the category of "age." In the great binary of youth and age, it is an understatement to say that in Western culture the favored of the two terms is youth. What implications does this have for technology? And technology for it? I am also interested in the relations between age and technology in general, or in what we might call the biotechnology of age in the culture of advanced capitalism. Most post-industrial societies are aging societies; that is, they are growing older in demographic terms, their populations are aging. In what follows,

I will speculate on these issues, focusing on the body and drawing on literature, film, and video.

As I have already suggested, the notion of technology as a prosthesis of the body underscores the extension of the body in space. But what of time? In a sense, one could say that many technologies—from the technologies of writing and speaking to those of film and photography—extend the body over time as well as in space. But these technologies produce only representations of the body. What about the extension of the organic body over time? If the new reproductive technologies can be said to destabilize power relations in our culture between male and female, can we conclude that new or emergent bio-technologies—I am thinking here primarily of longevity technologies—might destabilize dominant relations in our culture between the young, middle-aged, and old? Are there, in our culture, age-related prejudices associated with technology? Does our discourse about technology conceal an ideology of age?

In Gordon Rattray Taylor's mass-marketed paperback *The Biological Time Bomb*, the chapter bluntly entitled "Is Death Necessary?" contains section headings that transparently suggest some of our culture's concerns about prolongevity: "Eternal youth—allotted span—deep freeze—social consequences—arrested death—a friend in need—severed heads—immortality."[13] Note that the first phrase reads "eternal youth," not "eternal life" or "eternal age." It is clear that youth is to be valued; old age, not. Worse, old age—specifically, a long old age—is to be feared; there is an "allotted span," and we should hope not to live beyond it when death is "arrested," when death is our proverbial "friend in need" but does not come, when our bodies have withered away and all that is left of us is "severed heads."

[56] Taylor's reference to "severed heads" recalls science-fiction writer Olaf Stapledon's *Last and First Men* (1930). This ambivalent history of the future presents us with a stunning scene of giant human brains. Immortal but physically immobile, they thrive in beehive-shaped cells, nourished by a system of chemical plants. The implication? The desire for longevity results in the obliteration of the body of the species, and it is disgusting. Or consider an altogether different version of the future

of the body extending itself in time—a narrative not of the human species but, rather, of an individual in which the aging process is literally reversed. In a little-known short story by F. Scott Fitzgerald, a grown man reverts day by day to infancy; the dream of eternal youth turns into the nightmare of increasing infantilization. It would seem that when it comes to biological time, we cannot think of it in any way other than as an irreversible variable; that is one of the reasons why in the United States we cannot tolerate the possibility of people in their later years losing some of the abilities they have acquired over their life, reverting, as we say, to a "second childhood." (Many people have sex-change operations; can we imagine the real possibility of an age-change operation? I will turn to this later in a discussion of the film *Brazil*.)

In Aldous Huxley's *Brave New World* the ultimate fantasy of technological domination is over the human body, one achieved through biotechnology. Published over sixty years ago, this novel continues to speak to our society's dominant fantasies in relation to age today. In this new world, old age, defined as a problem of hideous proportions, has been conquered. All the physiological and psychological stigmata of old age have been eliminated. "Characters remain constant throughout a whole lifetime"; the body is programmed biologically to maintain youth and vitality over the long stretch of the life span—and then to die rapidly. As Huxley envisioned this "utopia," the process of aging and dying is telescoped into a single, short moment: "senility galloped so hard that it had no time to age the cheeks—only the heart and brain."[14] Thus, both the individual and society at large are conveniently spared any disquieting knowledge of physical aging. Moreover, the state is spared the troublesome task of having to decide whether to devote any resources to older, presumably "less productive" members of the population. There is no need for irritating debates on generational equity, for example. The problem has been solved with biotechnology. In consumer culture we have a name for this—planned obsolescence.

The head of state in *Brave New World* says, "We haven't any use for old things here."[15] For a society such as ours that is built on the technological values of efficiency, cost-control, and innovation (the "new"), what is perceived as "old" is understood as not only antithetical to our

[57]

dominant values, but dangerous. Our language in everyday life confirms that a technology that is "getting old" is suspect per se, as we see, for example, in this headline that appeared a few years ago in the *New York Times*: "Aging Nuclear Reactor Hazardous, Experts Say."[16] Every nuclear reactor is surely perilous, but setting this aside, what the content of the article makes clear is that the real problem is *not* in fact the age of the reactor but rather inadequate design—and, in particular, a design that is not Western or "modern."

Fundamentally, the term "aging" is derived from the organic realm. Aging is also a social process, and as such it is accorded different cultural values in different settings. The terms "youth" and "age" belong to the rhetoric of this continuum of the biological life span, which is given different meanings in different societies. But we also apply these terms to artifacts, as we just saw. And when we do, the negative connotations attached to the term "age" when it is used to describe, for example, a material object that is falling apart (such as a car or a washing machine), will return to the human domain laden with negative associations. We also use, of course, the terms "new" and "old" to describe both artifacts and organic beings—it's a "new" tree, we say, or a "new" car. But we do not say that a computer or a blender are "young." The problem, then, is that the term "old" but not the term "young" is transferable from the organic realm to the technological realm—and then back again. And generally speaking, in our technological culture there is nothing good about an artifact or a technology that is old. The supreme value is to work efficiently (which is understood to be synonymous with being new), not to break down. In short, I would argue that the rhetoric (as well as social practices) of the technological culture of advanced capitalism contributes to widespread ageism against older generations.

Age relations are power relations—with the young generally holding the power over the old (the young only need to hold on; in most cases, the passage of biological time will take care of any power differential in their favor). In Terry Gilliam's shrewd 1985 film *Brazil*, this relation between two generations is momentarily upset by technology; it is further complicated by gender. The setting is a bureaucratically impacted, grey retro world where one of the only vivacious beings is an "older" woman, a woman who is (of all taboo things) the mother of a grown son. With the aid of one plastic surgery after another, the mother (the beautiful Katherine Helmond) of the bureaucratically beleaguered Sam Lowry (Jonathan Pryce) appears to grow progressively younger. In a final scene she appears to be the youthful age of twenty-five, which unambivalently delights her even as it unsettles her irresolute son. Her plastic surgeon makes a million promises. "I'll make you twenty years younger," he says, "they won't know you when I finish with you." But another surgeon warns prophetically, "It will never last .... In six months she'll look like Grandma Moses." The younger generation would prefer the older generation to stay where they are—to stay older than *they* themselves are: it is a matter of power.

In *Brazil* nothing works efficiently. Actually nothing seems to work at all. One of the dominant institutions, Information Retrieval, is more than a frustrating joke, it becomes a bureaucratic site of torture. More importantly for my purposes, tampering with the biological body—particularly an older woman altering her own body—is implied to be not only unnatural, but worse, horribly doomed to fail. Katherine Helmond's temporarily youthful body is a postmodern parable for the injunction against biological engineering that would give an edge to the old. It is understood that she will be punished for her desire to claim the sexually admiring glances of men her son's age.

The disavowal of the limits of the body in technocriticism—or what I have been referring to as one of the ideologies of technology in general—is thus especially complex with regard to age. In *Brazil* what you want (and therefore what is "right") depends on just how old you are. In *Brazil* both the son and the mother want to escape the limits of the body. Both hope that technology will enable them to do so. The son tries to escape certain torture by fleeing in a car—actually he is only *imagining* the car. The playful, designing mother calls on the aid of surgical technology for her escape—actually, it is an *old-fashioned* technology (clumsy clamps are involved), so we as spectators, smugly ensconced in a technological future relative to the awkward technological culture of *Brazil*, know it won't work. In fact neither the mother nor the son

[59]

succeed in escaping. In one of the several fascinatingly grotesque scenes of the film, the son witnesses the future of his mother in the fate of one of her friends. This older woman has also had a series of plastic surgeries, and they begin to fail—that is, to reverse themselves. During her funeral Sam accidentally upsets her coffin—only to spill a disgusting gelatinous mass of decomposed matter that was once her body.

We also see ageism at work in the reactions reported recently to several new advances in reproductive technology. Consider the story about cloning that made the headlines in the United States and Europe in the fall of 1993. In a piece in the *New York Times* headed "Cloning Human Embryos: Furious Debate Erupts Over Ethics," several reasons were cited as to why the cloning of embryos is unacceptable—including, for example, that the sanctity of the individual would be violated, and that the human body should not be allowed to become a commodity. These reasons we could have easily predicted. But one objection in particular intrigues me. The director of the Center for Bioethics at the University of Minnesota, Arthur Caplan, raised the specter of age; the scenario of twins being born years apart was disturbing to him. " 'Twins that become twins separated by years or decades let us see things about our future that we don't want to,' Dr. Caplan said. 'You may not want to know, at 40, what you will look like at 60. And parents should not be looking at a baby and seeing the infant 20 years later in an other sibling.' "[17] Why not? Why should this be so potentially distressing, not to say intolerable? Why should a vision into our biological future be so disturbing? In our technological culture the future is associated with progress, with development, with the new. What is "aging" and "old" is considered anathema to it, associated with decline, diminishment, and disease.

[60]   Many object to certain high-tech reproductive procedures in principle for all women. But some only object to them for "old" women. The prejudice against older women is blatantly revealed, for example, in the horrified reactions to the reports of the "old" women—a fifty-nine-year-old English woman and a sixty-one-year-old Italian woman—who, with the aid of in-vitro fertilization, recently gave birth to children. This was considered sensational news. A January 1994 story

in *People* magazine, which was accompanied by full-page photographic spreads, carried the title "Turning the Clock Back." Many insisted vehemently that a postmenopausal woman should not have babies because it is "unnatural," not in the proper biological order of things. Obviously, however, what is also at stake is a deep-seated prejudice against older women; a pregnant older woman has long been understood as a monstrous contradiction in terms, both biologically and culturally. When the biological barriers are lifted, as they were in this case by technology, what remains so glaringly to be seen is the cultural bias against older women. In France, for instance, the government decided to introduce legislation to outlaw artificial insemination of women who are postmenopausal. As many were quick to point out, our culture does not condemn older men for fathering children; indeed older fathers are often seen as especially virile. So why should older women be vilified? What this story shows is that women are *aged by society* earlier than are men. What it also reveals is that if feminists have turned to the study of technology primarily through examining and championing reproductive rights for women, they have been conceptualizing technology as an issue for young and midlife women—not one for women of all ages.

We also see the ageism—and sexism—embedded in the ideology of technology in the reactions to the recently reported possibility of transplanting ovaries taken from fetuses into the bodies of women who are infertile. Again at stake is the presumed "natural" order of generations; many find the prospect of disrupting this "natural" biological sequence "grotesque," as a lawyer who is an ethicist at Harvard put it. "'Should we be creating children whose mother is a dead fetus? What do you tell a child? Your mother had to die so that you could exist?' George Annas asks."[18] The obvious implication of his objection is that a woman has to be a biological mother before she can be a grandmother. But why necessarily so? The proper focus, in my view, is on detaching biological roles and sequences from social roles, such as parenting, when it allows for increased opportunities to fulfill those roles.

Thus, on the one hand, I think we should explore in a positive spirit the possibilities offered by biotechnology as they intersect with issues concerning the aging body. Why should women not extend their biological capabilities with, for example, such reproductive technologies? On the other hand, the grandiose fantasy of eliminating altogether the inevitable process and consequences of aging reveals a prejudice against aging that is harmful to all of us. Indeed, the limits to the extension of the power and control of the human body with the aid of technology can be clearly seen in the context of the body as it ages.

Stanley Kubrick's elegant 1968 science-fiction film *2001: A Space Odyssey* begins, for example, as a technological romance with an ape who discovers in the midst of war that a bone multiplies the strength of his arm; he joyously hurls it into the air. The film then cuts to the future; we see a spaceship soaring aloft accompanied by the waltzing strains of the *Blue Danube*. The bone extends the power of the arm; the spaceship, the power of locomotion, of travel.

Kubrick's superb anthropological fable concludes, however, with a series of sobering scenes that remind us of the limit to this narrative of the technological evolution of the power of the prosthetic body—that limit is the aging body itself. The sole surviving member of the crew (Keir Dullea) has been captured by alien intelligence. Alone in an exquisitely tasteful, spare suite of rooms, he is witness, in a slow relay of looks, to his own future. It is the progressive aging of his body. He ages before his own eyes, soberly accepting this visual knowlege that Arthur Caplan (to whom I referred earlier in reference to the cloning of embryos) suggested we might not be able to acknowledge psychologically. Technology, it is thus implied, will never provide an escape from the upper reaches of the biological life span. (Earlier in *2001* the fantasy of prolongevity is also punctured when the computer HAL unplugs the life-support machines to which the hibernating crewmen are attached.) The sequence ends with the captain, frail and immobile, confined to his bed. But to this sequence Kubrick added a coda. A fetal figure of human life serves as a redemptive image, displacing aging onto the birth of life, recasting the narrative of technology as a mystery play in which the cycle of death and the resurrection of life is endlessly and magically repeated.

If we rewrite this epic narrative as a story of everyday life without the rescue of a coda, it might go like this: as children we take pleasure in progressively acquiring prosthetic expertise by learning how to use a fork, ride a bicycle, and operate a computer; as older adults we may welcome, although possibly ambivalently, the prosthetic innovations of bifocals and aids for hearing; but most of us fear the future prospect of frailty as a cyborg, "hooked up," as we commonly say, to a machine.

"The ultimate technological fantasy," Andreas Huyssen has written, "is creation without the mother."[19] This is a widespread male fantasy embedded within patriarchy, the fantasy that women—especially older women—are not necessary. But from the perspective of age, not gender, the ultimate technological fantasy is, as we saw in *Brave New World*, the perpetuation of the power, health, and beauty associated with youth coupled with the psychic assurance of immortality. Another way of making this point is by asking Donna Haraway the age of the cyborg she so optimistically celebrates. It is, I suspect, young, at the most middle-age. As I have been insisting, technological culture is a youth culture.

In video artist Cecelia Condit's magical miniopera *Not a Jealous Bone* (1987), we see a surprisingly enchanting reversal of the dominant values of youth and age. As in *2001*, the bone here too is a symbol of power—technological power, that is to say, cultural power. There are two main figures—the eighty-some-year-old heroine, Sophie, with dentures and an every-which-way black wig, and the conventionally beautiful younger woman, who is in, I would guess, her late twenties. In the struggle between them, it is Sophie who wrests the bone from the younger woman, who is as narcissistically preoccupied with her mirror image as the jealous Queen in *Snow White and the Seven Dwarfs*. It is a "magic bone," which represents power for the old in a technological culture fixated on youth.

When the video opens, we see Sophie making a spectacle of herself, her body ungainly and unsteady in its old-fashioned, black-skirted bathing suit. At the narrative's end, Sophie, now the possessor of the bone, dresses up with charming pleasure. No longer culturally reprehensible, she dances lightly and with self-possession on a center stage. Moreover, she tosses the bone into the unseen audience, handing it over to others who would catch it. It is the desire for life, not youth or the appearance of it, that is given value—in the figure of an older woman.

In *The Biological Time Bomb* Taylor warned of the dangers of pro-longevity for capitalist culture. Who does he single out as representing a sure threat? Older women. Consider this passage, which is rife with both sexism and ageism: "Another probable consequence of the prolongation of life would be a further increase in the excess of women over men in the higher age groups of the population .... A society containing a large percentage of such 'dragons' would be intolerable."[20] The eighty-some-year-old Sophie dancing on the beach, her makeup askew and her body unseemly by our youthful standards: Taylor would cast her as a *dragon*, as a *biological time bomb* that is threatening the vitality of our culture. Taylor would want to stop her clock for his own purposes.

In *The Biological Time Bomb* Taylor also worried that a significant increase in the older portion of the population relative to those younger "might bring about a gerontocracy and an 'age-centered' society; it is difficult to see how the latter kind could be as vigorous as the former."[21] In terms of the rhetoric of age, we see here again the metaphorical use of the term "age." But here the rhetoric of biological age—"young," "aging," and "old"—is applied not to an individual or to an artifact but rather to an entire society. Implicit in the notion of an aging society is the fear that it—therefore we—might die. I read the marvelous figure of Sophie as a protest against the youth-oriented—and sexist—ethic of our capitalist culture. As our population ages—in part, ironically, because of technological change of many kinds—the United States has grown anxious that it is losing its competitive edge to "younger" economies in Asia, a fear that only works to further intensify the ageism inherent in an "advanced" technological society.

Homo Generator:
Media and Postmodern Technology

Wolfgang Schirmacher

I. Introduction

What is real? How do we know the truth? Today, these basic philosophical questions are harder than ever to deal with. Even if we trust the
lessons of daily life, it is difficult enough to come up with convincing
answers. For students in rural Pennsylvania, nature may still seem very
real; for the kids in the big cities hooked on Nintendo, "nature" is the
name of a computer game. And the video and cable TV addicts around
the globe—from the farmers in India to the unemployed former Party
officials in Moscow to the graphic designers in New York—leave the
misery of a so-called reality behind them and embrace true life in the
movies and soaps. Are they merely escapists, or are they citizens of a
new world civilization that most of us still violently deny?

Let's have a closer look at the other end of the social ladder:
Executives and administrators are dependent on statistics generated by
information technology. For them there are no goods or customers or
students, the bottom line is numbers, the sole judge of their performance. How real is politics these days? The media is the racetrack,
polling tells us what we are betting, and success is measured in soundbites and news outputs.

Most people over sixteen pride themselves on being able to easily
distinguish between natural and artificial reality, but their own actions
will often betray them. It's ten o'clock—do you know which program
you are in? Self-criticism, humanity's sharpest weapon, advises us to
ask a tougher question: How can we know what is real? Philosophy,
science, even common sense, appear flawed, biased, and tainted by
instrumental reason. In the postmodern age no authority seems to be
left to tell us what to believe. Of course, we still have ourselves and that

means thinking power, intellectual honesty, intuitive understanding. It looks as though we can know only what we ourselves discovered as witnesses to a truth lived personally. We are victims and victimizers of life and, therefore, maybe not the most trustworthy witnesses (ask animals and plants about it), but I am the only me I've got. What we as creators, participants, and observers can know about today's reality points to an immense shift: the emergence of artificial life as the reality for human beings. Artificial life is an anthropological fact as well as a challenge being met in the two most exciting (or frightening) developments in science, technology, and society: biotechnology and communication technology. If biotechnology provides the hardware of artificial life, communication technology designs the software, which includes a postmodern culture. It is within this cultural environment that we decide how we should act, what we hope for, and, finally, what it means to be a human being.

But shouldn't we be more careful in introducing a concept as paradoxical as "artificial life"? Can "artificial" in respect to life be more than merely a technocratic metaphor? Life seems to be the one thing over which we have no control, a force of the universe rather than a human enterprise. Life can be simulated artificially, not created by us. Steven Levy, in his book *Artificial Life: The Quest for a New Creation*, disagrees and challenges us to give up the distinction between simulation and reality.[1] If life is characterized by the ability to create something out of nothing, as well as the ability to change and custom-tailor itself to its environment, then self-replicating computer programs constitute life, Levy argues. The radical new computer-science technique sounds familiar: evolution. But this is exactly the reason why we should be suspicious; in duplicating a view of life which is itself a nineteenth-century concept of trial and error and survival of the fittest, we only prove ourselves anachronistic. To condemn artificial life to the fate to which we've delivered natural life would be a real folly in light of the ecological crisis. Obviously, "artificial life" is a concept that—for better or worse—seems to mean "substitution of nature." In my view, this is completely misleading, a false understanding that has no basis in reality. An understanding of artificial life as substitution of nature promises what it can't deliver: a brave new world where technology rules supreme.

Artificial intelligence (AI) once incorporated one of these big promises—the vision of abolishing all constraints on human intelligence in a next step of evolution—but has since been humbled. Now it's just AI still struggling with all the basics of sensing and speaking. Computers may be able to do incredible things, but it is not likely that they can serve as a paradigm for intelligence—the computer metaphor is hopelessly overstated, as John Searle points out in *The Rediscovery of the Mind*. Humans don't compute, they understand intentionally.[2] And the cyborgs, those hybrids of humans and machines, have turned out to be truly at home in the virtual reality of artistic adventures.[3] The brain and its artificial body, the body and its artificial brain will be considered "inhuman" when they appear.[4] (So let the scientists do it to their kids first!) But realized by biotechnology and communication technology alike, artificial life could overcome its genesis in technocratic ideology and mend the flaws of its anthropocentric intentions. "Artificial life" is a prospect which should enable humanity to attain its own fulfillment instead of triggering our resistance.

Are we using the wrong terms? According to William Safire, language watchdog of the *New York Times*, the once influential term "artificial" has been dismissed because it means "fake" in English, and "virtual" has become the new buzzword because it reads "almost." This would certainly meet the much lesser expectations we have in this field nowadays: "almost" life, "almost" reality, "almost" intelligence would leave our traditional worldview intact, merely adding new layers to it. "Virtual reality" simulates reality by creating a "double world" in which new possibilities may be explored. Such a term is especially useful in hammering out a sharp distinction to "artificial life" and discarding the misleading "fake" analogy. "Artificial life" is human-generated and shares this characteristic with "fake"—both are artifacts. But "artificial life" doesn't imitate life, has no original, doesn't work with the blueprint of nature, and is, as such, completely different from "fake." Instead, artificial life describes the only life we know anything about: humanity. To call this reality a fake, a reality which shows itself as the human condition of natality and mortality, as *Dasein*, or being-in-the-world, is a flagrantly ideological and biased statement. Artificial life embraces the art of living unique to humans in giving birth to the unexpected (Hannah Arendt's natality) and releasing the event of death (Martin Heidegger's mortality). Artificial life is effected in life technologies which range from breathing to computer hacking. There is no anthropological difference between watching television and smiling at another person—artificial life celebrates itself in both techniques.

II. The Artificial Body of Media: Homo Generator at Work

> Wahrheit "gibt es" nur, sofern und solange Dasein ist.
> —Martin Heidegger, *Sein und Zeit*

Truth is a gift of *Dasein*, which is our place and activity in the world's process. As artificial beings by nature, our body as well as our mind is "a happening of truth at work"[5] every lived-through moment. Even in the most inhumane enterprises, truth is still at work in humanity as a silent cry toward its absence. According to Heidegger, truth is not about being right or wrong but accepting "aletheia,"[6] the powerful interplay of revealing and concealing, which shapes humanity's destiny. Today's media, taken as one of our body's splendid incarnations, is involved in the interplay of instrumental technology and life technology, torn between the murder of the body and its elevation (Hegel's *Aufheben*) to an unknown status.

If we take a critical look, what has modern technology done to our bodies? Since the nineteenth century a person's authentic "home in the world" (Maurice Merleau-Ponty) has been turned into a machine, serving as an object alienated from itself as origin. The revolution of the artifacts is perceived as a negation of the natural body. Dolls function as an inscription of the body, bearing our consent to the abstraction game that has been going on for quite some time now. Our culture is fascinated by the immaterial body which knows no aging process and may overcome even death. Modern technology seems to be as hostile to the natural body as premodern Christianity. The lesson of Frankenstein is

[68]

lost on a society using plastic surgery more and more frequently to reshape a body and in which scientists would love to keep a brain alive without a body. Donna Haraway's cyborgs are around the corner, bio-robots which cannot die, with bodies that will not age (only rot, I hope). In virtual reality we could be with Jesus Christ in Golgotha or with Marilyn Monroe, any day, no stinking sweat, no salty tears, no inner-body noise—only the s(t)imulation of it. Today's media are a prime example of a politics of the body which is basically applied technology and has crossed the last frontier, the natural body. Even food, satisfying the single most natural urge, increasingly represents a deep rejection of the body, as it becomes cyborg food on our TVs.

Or do we misunderstand what's really going on, is our perception biased, unable to free itself from the ills of authority, anthropocentrism, and logocentrism? This traditional view of the world functions as a useless filter, darkened by truism as well as by scientific knowledge. Can't we take hope in a fundamental shift to be observed from the industrial to the postindustrial, from hardware to software, from seriousness to play?

It seems obvious that the media is a body-machine that no longer belongs to a modern framework of the grand narratives of emancipation and progress[7] but, rather, to the trivial postmodern tales of "everybody is an artist" and "anything goes." Postmodern communication technology provides opportunity for mass participation and blurs the distinction between creator and audience. New media forms such as camcorders and computers compose and design systems that allow the freeing of creativity, appearing to contradict Martin Heidegger's and Theodor W. Adorno's predictions that modern technology will transform creative uniqueness and artistic innovation into a ready-to-consume item of a giant supermarket called "culture."

Or are we only fooling ourselves? Does the simulation of creativity in communication technology challenge a negative dialectics of technology: death technologies versus life technologies, being mediocre versus becoming oneself, participating versus originating? The ideological uses of these contradictions are widespread, especially in the form of body versus virtual body, reality versus virtual reality, true versus false.

The construct of dichotomy was helpful once in differentiating between friends and enemies when formulating the political correctness of his or her own viewpoint. Contemporary critics of communication technology such as Neil Postman, Stuart Ewen, and Paul Virilio have addressed homo faber—the artist as engineer—and his or her product of a fast-food media and kitsch, but have failed to recognize what I call "homo generator," the media artist as generator of human reality and his or her responsibility for tomorrow's artificial world. Rap musicians' and hip-hop techno-aesthetes' reconstruction of black culture through living technologically has long been overlooked. The "radical technologists" may still believe that they use technology before they are used by it, but this stems from a philosophical ignorance that ill suits their technological smartness. All these new radicals are motivated by an overwhelming need to take responsibility for the rotten state of the earth—and rightly so.

Homo generator realizes the hope and the angst of the post-Hegelian philosophers, a *Dasein* beyond metaphysics, a human being that needs no Being, no certainty, no truth.[8] Modern technology is the birthplace of homo generator—but an ambiguous one. Homo generator is not a telematic tale of technological triumph over nature, as Vilém Flusser used to tell it (before he died in a car accident in 1991), nor proof of a technological mutation to a higher form. Homo generator is not a success story and will not vouch for progress, but by the same token homo generator begins to fulfill the artificial existence of humanity. Jean Baudrillard analyzed this phenomenon in communication when he wrote: "The only irresistible pulse is to occupy the nonspace, the empty space of representation that is, par excellence, the screen."[9]

Yet homo generator in my understanding revokes a misconception of humanity. Homo sapiens, homo faber, homo creator—none touches our core any longer. We shall, of course, keep on thinking, continue to make tools, and imagine ourselves to be the originator of things, of our acts and thoughts. But all this pales in comparison with the immense ability to produce new forms of life and determine the biological as well as the spiritual future of the earth. This homo generator is no longer a gifted dilettante whose successes are owed principally to chance. Homo generator does not have to settle for what's given;

he or she works instead, without any restrictions, with the fundamental building of life in all forms.

Whether or not we accept the fact, we have become homo generator, most visibly in gene technology but equally so in communication technology. A philosophical challenge without precedent lies in discovering and reinforcing those traits already in accord with this new type of being. We could not have discovered gene technology or virtual reality as a human capability if it did not belong to our nature and if it were not a characteristic of our life technology. If we understand homo generator not as an extreme case but as the norm, one of our conjectures about humanness is then strengthened, one which has long determined us subconsciously but is uttered only with great aversion. If we penetrate all the dissemblance and tear away the last veil of analogy between human and animal, it then becomes irrefutable: we are but artificial beings among all other beings, our bodies are artifacts by nature.

Homo generator's body politics is to see/hear/smell/touch/taste/think before you act, it claims aesthetic perception as the basis of comprehending and interaction. Homo generator has no fear of his or her mistakes, for they are inseparable from his or her succeeding—as body politics teaches us. Responsibility also means being able to assume one's guilt and to reject blame for anything you have not caused yourself. Homo generator is a substantial beginning, unique but not original, self-care without egotism.

How do we recognize such a fabled being? Could we find homo generator as one of the high priests in the labs at Rockefeller University? Do the superstars of today's media follow the model of homo generator? Or is a computer program more likely to represent what is challenging in human existence? Is homo generator beyond the gender issue? It is much easier to tell what homo generator is not (not a scientist, not Madonna, not cyberpunk, not gender neutral) than to indicate how he or she may appear. I would like to follow "traces" (in Jacques Derrida's sense) beyond simulations in postmodern communication technologies.

There is no revealing without concealing, as the poets have long demonstrated. Revealing deconstructs, opens up, tears the fabric of the known. Revealing through media brings back the body in amazement, in laughter, and even in disgust. Communication technology provides the software for generating the body, being yourself, overcoming self-censorship. As the Beastie Boys put it: "You have to fight for the right to party!" "Death technologies" is a corporate identity!

Isn't that a great artificial life? Icons in politics and pop culture—media can force them to dance! Laughing in the face of death, as

We will, we will rock you.

Hello, my name is Ann Cook and I would like to ask Mr. Gorbachev how Raisa is.

I'm not going to kill you, but I'm going to tell you three things that will haunt you for the rest of your days: You've ruined your father, you've crippled your family, and baldness is hereditary.

Georges Bataille teaches us, overthrows common sense and transgresses decency. Homo generator is rebellious, takes no prisoners, interrupts quite violently the daily routine. But all that with a smile, please. Postmodern irony is the style of homo generator, generating a rich texture of codes and speaking at least two languages at the same time. The disclosure of truth is never overt but, rather, an indirect communication and a complex happening in life that needs concealing as much as revealing. Difference results from the power that lies in the unknown, in the not-yet-known or never-known. Absence prevents anthropocentrism, the greatest threat to humanity. Ambiguity returns to us the insight, the intuition our bodies possess. Difference, absence, ambiguity: these modes of being belong to the realm of concealing, the most compelling life technology we can encounter. The ongoing process of concealing protects homo generator from participating in, buying into, being part of what others have revealed and imposed upon us as "given." This addresses gender roles as well as schools of thought, national identities as well as traditions.

Concealing has the saving power of disengaging us from this phony givenness and connecting us with the body as our artificial "home in the world." Therefore, in media we fight for our right *to know* as much as for our right *to know that we don't know* (not to be confused with ignorance). Traces of nonpresence, the interplay of present and absent, the disturbance of the sublime, and the transcendence of terror indicate a concealing which will never reveal itself. Innovations in communication draw from the hidden whose influence cannot be overestimated. Such powerful concealing allows us a postmodern technology which has abandoned the question of control. It is the Socratic power of "eros"—that which we don't have, see, know. Nobody's master and nobody's slave, communication technology enjoys concealing the obvious, the daily routine as much as obsession. We are how we eat, to what we aspire, and for whom we feel compassion. In artificial life we communicate the way other species breathe. Blaming media and technology for our cowardice and for our fear of being unable to live artificially with grace is self-defeating. Innovations in communication are judged solely by their potential to let us be artists of living, working on a social sculpture as Joseph Beuys envisaged.[10]

Eat dessert first! We eat in order to survive but we don't eat dessert in order to survive. We eat dessert in order to get a feeling of what life is about.

I ate oatmeal at about 6:30 this morning, and kukicha tea, and I had an egg salad sandwich in the middle of rehearsal, on rye toast, and I just ate some walnuts and sunflowers seeds.

You know what they can do for us? They can suck our motherfucking dicks!

It is no good to tell humanity to stop fighting for survival—in these fights all our best features are in.

III. Media and Postmodern Technology: Ironic Laws

It has become increasingly rare to communicate without the help of media-generated forms, role models, and channels. Seeing and hearing, which inform our thinking, are overwhelmed by communication technology. Thinking and writing are done within a framework provided by media and follow the possibilities and limits of the chosen technology. Two examples: The telephone answering machine has changed our style of personal communication as lastingly as the computer changed the flow of writing. With the feature "call screening" we are present and absent at the same time, becoming free and open to choose the "essential moment," as the Zen masters put it. And the ease with which we can alter and expand the text already written into the computer eliminates the "voice of authority" in us and allows us to play with thoughts. Of course, one could describe both activities very differently, calling the answering machine a pest and the end of meaningful communication and word processors the source of text pollution as they tempt even the serious person to babble on like a child. Few people feel excited about the prospects of an artificial life shaped by communication technology; most feel disembodied. But how we judge this impact is less important than acknowledging how the fundamental changes in our lifeworld through media are taking form.

A valid media criticism should be based on media aesthetics, the judgment founded on experience. Instead, experiences with media have been held up to a value system not intrinsic to communication technology itself. What is missing is an unbiased sensual experience with media, a round-the-clock treat with film and video, sound studio and computer graphics, multiple television channels and a working remote control. Missing is living with fax and laser printer, with notebook computers and High-Definition Television (HDTV), being at home in computer conferencing, missing are legal hackers riding the waves of cyberspace. Hyperreality and virtual reality are concepts that up to now have only pretended to have experienced what they describe. To experience a phenomenon aesthetically we share in it, follow its own style. An account of this exploration should describe the phenomenon in its own terms. Living media is most adequately experienced in the industrialized countries, where it is a universal phenomenon, not a life style of the happy few. The noise and crime, the homeless, the cars, and the garbage of Manhattan are not contradictions of but, rather, complements to the artificial life style media demonstrates so convincingly. It's all human-made, the bad, the ugly, and the beautiful, but no longer controlled by humans. New York, the media center of the world, designs an art of living which fails as often as it succeeds.

In terms of ontology, it is the human contribution to Being. But no one ever intended it to be this way: we don't want to participate in performance art, each one of us is simply fighting for survival. Georg Wilhelm Friedrich Hegel's insight into the unintentional event which he termed "cunning of reason" might be helpful in understanding media.[11] But we then have to sidestep Hegel's belief in progress and should anticipate, following Arthur Schopenhauer, that this art of living may lead to the infamous art of torture.[12]

What was once called mass media has radically changed in meaning. Instead of media for the masses, we can observe the birth of a potential media *of* the masses. Only with regard to distribution is "mass" still a topic—in terms of audience, we are back to tailor-made programming and inventions. This artificial lifeworld reveals laws that follow the "open-system irony" described by Richard Rorty.[13] Such irony is an observation, not an opinion, and allows for a polyphonic language of several codes. Sören Kierkegaard called it a "mastered irony" that doesn't want to shock or provoke but expresses the sense of a break.[14] There is no final word in the artificial lifeworld, and homo generator—"being him or herself, only more so"—is used to indirect communication. We have to take the ironic laws of media evident in a postmodern world both seriously and not seriously. Aesthetics simultaneously perceive success and failure in communication technology and experience media as our body living playfully in oppositions.

The first law of media is: *The self is the focal point.* This self is not the ego of domination or the subject of modern times but the activity of "caring for one's self." Taking care of oneself is now the activity of media. In advertising, to use a telling example, the message remains the same in any production: "You, only better!" Advertising, indeed all media, addresses the "Ideal Ego" (Jacques Lacan), the primary unity we lost with the mirror phase and through the distraction of language. The promise of mediation in media hints at a possible reunion with the world. The self is in no way satisfied by being apart and single. The self wants to overcome its separateness without losing its specialness. The self of media is an activity. In media we are confronted with the challenge of writing our own lives—with camcorders as well as with computers, with answering machines as well as with films. Mouse and remote control are only the beginning of interactive features in media that allow us to edit and cut, stop and go, break and let flow whatever situation we encounter. In media we write our autobiography—and if we don't, somebody else will do it for us. The care for self is a project, not a given fact, and its other side is neglect.

What can a self, which by definition doesn't care about having a fixed outlook and a secure place in society do? It has to become a creative self. Such a self is the nightmare of traditional media hype and the dream of innovative media producers. This creative self is elusive, interrupting the conventions of dominant culture by twisting it around. Bart Simpson, as well as Beavis and Butthead, are the icc   . this style of subversive irony, which mocks even the most serious things. Gone is the archaic dichotomy of form and content that reinforced the taboo of seriousness.

Performance is the signature of truth. This second law of media asks you to sign your name to the event, with no credit given for the hidden agenda. Media babies nurtured by shows, soaps, and trash movies live happily with collage, parody, and pastiche characteristic of the realm of performance. "Tele-Vision is Tele-Action" (Paul Virilio). There is no dialogue outside media, and all the action takes place within. Truth is hard work, done with a smile. No media event has the authority to enforce action—propaganda's heyday is over. Media has to seduce and open up a field of action which has no goal other than playing life, re-arranging a never fixed lifeworld. Media's seduction has nothing to do with the so-called power of pictures: media seduces by style alone.

Style is the medium of action. This third law of media addresses the "deframing" of our perceptions. Style is a self-evolving activity produc-ing a gaze and opening the ear. It is not the author's viewpoint, or his or her aesthetic judgment that style expresses. Style is a game playing with time and language in which you discover and forget the self. Style is neither an identification tag nor a tool of power but a composition never made before, in a language free of fixed meaning but still mean-ingful to you. Media needs style in order to resist the displacement of creative production through mechanization. In today's highly tech-nological media the machine's mood of standardization is restricted to machines.

Through style alone the self overcomes the hidden danger that attends every use of technology. What you call your tools (or toys, at best) and for which you claim "total control" is, in fact, shaping your imagination and limiting your options. New media personalities tend to be cyborgs without knowing it, locked into their multimedia pro-grams in joyful bliss. They seem to have forgotten that until now com-puters had no sense of beauty: "Whatever a computer draws in graphics looks like shit" (Mitchell Feigenbaum)—unless it is copied from a human design.

But wouldn't this prove the tool status of computers? Are their limi-tations not the best evidence of our control? In today's media world it's hard to tell who is the controller and who is controlled, and after a while it no longer matters: instead of ecstasy and intensity, the styles of the self and multimedia express connectivity and application, the skills of an instrumental lifeworld.

And more bad news is ahead: Professor Feigenbaum of Rockefeller University, one of the famous chaos theorists, finally taught his computer how to draw a beautiful curve entirely by itself. It took Feigenbaum more than two years, and the result is a glimpse of true computer style you may admire in the first *map of the world* drawn by computers.[15]

Style is openness, a life search that *"de-appropriates"* (Avital Ronell) given realities. The gaze as exchange of seeing and being seen, the ear as confluence of hearing and being heard, and writing as expression and learning do not need recording but aspire to fulfillment. Media fulfills itself in mediation, which has no outside goal. This mediation without goals rejects synthesis and embraces life games artists play: acting without believing, waiting without expectations, living without the will to survive.

The fourth law of media states: *Mediation is the flow of media.*

Mediation is no longer a deal between partners or a communication following established rules, but an innovative process of media to which we belong. In such a mediation there is not even the goal of mutual understanding, because the flow needs breaks. Dissent is the salt of mediation and designed to eliminate anthropocentric arrangements, the mafia practices of humankind. Mediation floods any content, fills the artificial lifeworld, evokes the "fourfold" (*Geviert*),[16] and allows us to be life's own artists.

Notes

Stanley Aronowitz

1. Donna J. Haraway, "A Cyborg Manifesto: Science, Technology, and Socialist-Feminism in the Late Twentieth Century," in *Simians, Cyborgs, and Women: The Reinvention of Nature* (New York: Routledge, 1991), p. 149.
2. Ibid., p. 150.
3. Howard Rheingold, *Virtual Reality: The Revolutionary Technology of Computer-Generated Artificial Worlds—and How It Promises to Transform Society* (New York: Simon & Schuster, 1991), p. 345.
4. Ibid., p. 346.
5. Shoshana Zuboff, *In the Age of the Smart Machine: The Future of Work and Power* (New York: Basic Books, 1988).

Herbert I. Schiller

1. *1991 Economic Report of the President* (Washington, D.C.: GPO, 1991), p. 303.
2. Robert Coen, "Vast U.S. and Worldwide Ad Expenditures Expected: The Next 20 Years in Advertising and Marketing," *Advertising Age* 5, no. 49 (November 13, 1980), pp. 10, 13.
3. Ibid., p. 10.
4. Herbert I. Schiller, *Mass Communications and American Empire* (Boulder, Colo.: Westview Press, 1992).
5. William J. Broad, "Swords Have Been Sheathed But Plowshares Lack Design," *New York Times* 5 February 1992, p. A1.
6. Herbert I. Schiller, "Industrial Light and Magic," in *Culture, Technology and Creativity in the Late Twentieth Century*, ed. Philip Hayward (London: J. Libbey, 1990), p. 128.
7. Stephen Holden, "Strike the Pose: When the Music Is Skin-Deep," *New York Times* 5 August 1990.
8. Martha Rosler, "Image Simulations, Computer Manipulations, Some Considerations," *AfterImage* 17, no. 4 (November 1989), p. 10.
9. Mark Crispin Miller, *Boxed In: The Culture of TV* (Evanston, Ill.: Northwestern University Press, 1988).
10. "How MTV Has Rocked Television Commercials," *New York Times* 9 October 1989.
11. Ibid.
12. Karen DeWitt, "MTV Puts the Campaign on Fast Forward," *New York Times* 8 February 1982.
13. Albert Scardino, "TV's Pace and the Ads Increase as Time Goes By," *New York Times* 11 September 1989.
14. Michael Wilmington, "Point Break: The Surf Is Up But the Credibility Is Down," *Los Angeles Times* 2 July 1991.
15. Michael Wilmington, "Science Runs Amok in 'Lawnmower Man,'" *Los Angeles Times* 6 March 1992.
16. Rone Tempest, "Euro-TV Tunes In To Hollywood," *Los Angeles Times* 11 July 1991.
17. Stewart Toy, "Mouse Fever Is About to Strike Europe," *Business Week* 30 March 1992, p. 32. Actually, Euro-Disney, to date (early 1994), has been a commercial disaster.
18. Jonathan Klein, quoted in *Culture, Technology and Creativity*, ed. Hayward, p. 138.
19. Leigh Bruce, "Europeans Tune in to a Wave of Television," *International Herald Tribune* 11 June 1991.
20. See Neil Postman, *Technopoly: The Surrender of Culture to Technology* (New York: Knopf, 1992).
21. Robert Kubey and Mihaly Csikszentmihalyi, *Television and the Quality of Life: How Viewing Shapes Everyday Experience* (Hillsdale, N.J.: Lawrence Erlbaum Associates, 1990), p. 189.
22. The Morelia Symposium Declaration, "Approaching the Year 2000," reprinted as an advertisement in the *New York Times* 1 November 1991. Apparently, the newspaper did not consider the declaration a newsworthy item.

Kathleen Woodward

1. William Kuhns, *The Post-Industrial Prophets: Interpretations of Technology* (New York: Weybright & Talley, 1971).
2. Andrew Ross, *Strange Weather: Culture, Science, and Technology in the Age of Limits* (London: Verso, 1991), pp. 9, 88.
3. See Michel Foucault, *The History of Sexuality*, Vol. 1: *An Introduction*, trans. Robert Hurley (New York: Vintage Books, 1980).
4. Marshall McLuhan, *Understanding Media: The Extensions of Man* (New York: New American Library, 1964), p. 19.
5. Hans Jonas, *Philosophical Essays: From*

Ancient Creed to Technological Man (Englewood Cliffs, N.J.: Prentice Hall, 1974), p. 78.

6. Sigmund Freud, *Civilization and Its Discontents* (1930), *The Standard Edition of the Complete Psychological Works of Sigmund Freud*, trans. and ed. James Strachey, 24 vols. (London: Hogarth and Institute of Psycho-Analysis, 1953–74) 21, pp. 90, 91–92.

7. Allucquére Roseanne Stone, in an essay on the body and virtual reality, astutely comments: "The quality of mutability that virtual interaction promises is expressed as a sense of dizzying, exciting physical movement occurring within a phantasmic space—again an experimental mode psychoanalytically linked to primal experiences. It is no wonder, then, that inhabitants of virtual systems seem to experience a sense of longing for a space that is simultaneously embodied and imaginary, such as cyberspace suggests. This longing is frequently accompanied by a desire, inarticulately expressed, to penetrate the interface and merge with the system" in "Virtual Systems," *Incorporations*, ed. Jonathan Crary and Sanford Kwinter (New York: Zone Books, 1992), p. 619.

8. Much has appeared in the popular press, however. *Time* ran a cover story in November 1993 entitled "Cloning Humans," for example, and a major essay entitled "The Genetic Revolution" and authored by Philip Elmer-Dewitt appeared in the January 17, 1994, issue, summarizing the revolution quite competently as follows: "The ability to manipulate genes—in animals and plants, as well as humans—could eventually change everything: what we eat, what we wear, how we live, how we die and how we see ourselves in relation to our fate," p. 42.

9. I am aware, of course, that there is a goodly amount of scholarly attention devoted to many issues in bioethics; in the university, however, courses on bioethics are often taught only in medical or nursing schools.

10. Shulamith Firestone, *The Dialectic of Sex: The Case for Feminist Revolution* (New York: Bantam, 1971), p. 228. See Valie Export, "The Real and Its Double: The Body," *Discourse* 11, no. 1 (Fall–Winter 1988–89): pp. 3–27. For a persuasive argument that the point of Firestone's book is "to stimulate a

revolutionary consciousness here and now rather than to dictate how babies should be made," see Carl Hedman, "The Artificial Womb—Patriarchal Bone or Technological Blessing," *Radical Philosophy* 56 (Autumn 1990), p. 19. Much of the research by academic feminists regarding reproductive technology is conservative. Shelly Minden's point is representative; she writes, "Although these technologies promise things that many women want—possibilities of healthier babies and of reduced infertility—the price that they exact is no less than that of women's autonomies over their own bodies," in "Patriarchal Designs: The Genetic Engineering of Human Embryos," in *Made to Order: The Myth of Reproductive and Genetic Progress*, ed. Patricia Spallone and Deborah Lynn Steinberg (Oxford: Pergamon Press, 1987), pp. 102–103.

11. Richard M. Restak, *Pre-Meditated Man: Bioethics and the Control of Future Human Life* (New York: Penguin Books, 1977), p. 167.

12. Donna J. Haraway, "A Cyborg Manifesto: Science, Technology, and Socialist-Feminism in the Late Twentieth Century," in Haraway, *Simians, Cyborgs, and Women: The Reinvention of Nature* (New York: Routledge, 1991), p. 161.

13. Gordon Rattray Taylor, *The Biological Time Bomb* (New York: North American Library, 1968).

14. Aldous Huxley, *Brave New World* (New York: Harper & Row, 1969), pp. 37, 135.

15. Ibid., p. 148.

16. The story appeared in the *New York Times* 25 March 1992.

17. Gina Kolata, "Cloning Human Embryos: Furious Debate Erupts Over Ethics," *New York Times* 26 October 1993.

18. Gina Kolata, "Fetal Ovary Transplant Is Envisioned," *New York Times* 6 January 1994.

19. Andreas Huyssen, *After the Great Divide: Modernism, Mass Culture, Postmodernism* (Bloomington: Indiana University Press, 1986), p. 70.

20. Taylor, p. 115.

21. Ibid., p. 115.

Wolfgang Schirmacher

1. Steven Levy, *Artificial Life: The Quest for a New Creation* (New York: Vintage Books, 1992).

2. John R. Searle, *The Rediscovery of the Mind* (Cambridge, Mass.: MIT Press, 1992).

3. See "Cyborgs at Large: Interview with Donna Haraway," in *Technoculture*, ed. Constance Penley and Andrew Ross (Minneapolis: University of Minnesota Press, 1991), pp. 1–26.

4. See Jean-François Lyotard, *The Inhuman* (Stanford: Stanford University Press, 1991).

5. Martin Heidegger, "The Origin of the Work of Art," *Poetry, Language, Thought*, trans. Albert Hofstadter (New York: Harper & Row, 1971), p. 36.

6. Martin Heidegger, *Aletheia. Vorträge und Aufsätze III* (Pfullingen 1967).

7. See Jean-François Lyotard, *The Postmodern Condition* (Minneapolis: University of Minnesota Press, 1984), pp. 31–37.

8. See Wolfgang Schirmacher, *Technik und Gelassenheit* (Freiburg/München: Alber-Verlag, 1983); (Ereignis Technik: Wien: Passagen-Verlag, 1990).

9. Jean Baudrillard, "The Télécratie and the Revolution," *Newsline* (March–April 1992), p. 1.

10. See Lutz Mommartz's video, *Joseph Beuys: Social Sculpture* (11 min, 1969).

11. G. W. F. Hegel, *Encyclopedia of the Philosophical Sciences* in *Outline and Critical Writings*, ed. Ernst Behler (New York: Continuum Publishing, 1990), p. 126.

12. Arthur Schopenhauer, *The World as Will and Representation*, vol. 2, trans. E. F. J. Payne (New York: Dover Publications, 1966), p. 581.

13. Richard Rorty, *Contingency, Irony, and Solidarity* (Cambridge: Cambridge University Press, 1989), p. 73.

14. Sören Kierkegaard, *The Concept of Irony* (Princeton: Princeton University Press 1989).

15. *Hammond Atlas of the World* (Maplewood: Hammond Inc., 1992).

16. *Geviert*, German for "squared," is a poetic metaphor. For Friedrich Hölderlin's description of the "fourfold" elements of a comprehensive understanding of human existence: heaven, earth, gods, and mortals, see Martin Heidegger's "The Thing," in *Poetry, Language, Thought*, ed. Albert Hofstadter (New York: Harper & Row, 1971). A poster-installation "Philosopher-artist's *GEVIERT*" has been my contribution to the exhibition "The Rape of Language" (Kunstverein Hofgarten, Munich, Germany, in May 1993). See Wolfgang Schirmacher, "The City as *Geviert*: Questions Arising from a Philosophy of Architecture, Sentence Building," *Architecture-Theory-Criticism*, ed. N. Hellmayr (Freyberg: Graz 1993).

II

Technology and the Body: The Constitution of Identity

The Merging of Bodies and Artifacts in the Social Contract

Elaine Scarry

Consent—across such separate spheres as medicine, political philosophy, and marriage law—has a stable set of attributes. The first and most essential of these, as I have argued elsewhere, is the grounding of consent in the body, the grounding in the material world. As one moves across the classic textual elaborations of consent—Plato's *Crito*, John Locke's "Second Treatise on Government," Rousseau's *Social Contract*, canonical medical case law such as *Schloendorf* v. *Society of New York Hospital*, and both religious and secular accounts of marriage contract— what becomes clear is that the body is, first, the thing protected by consent; second, the lever across which rights are generated and political self-authorization is achieved; third, the agent and expression of consent and hence the site of the performative; and fourth, and finally, a ratifying power. Together, these features make it clear that the consensual is never an unanchored act of volition: it is structured by its location within the material world.[1]

This conflation with the body is, then, a first essential attribute of consent. Yet no sooner is consent situated in the body than a second transformation takes place: once the volitional is located in the material, the material is now itself reconceived as artifactual; it is perceived as a made thing and described in the idiom of creation or invention. This two-part structure of perception—the location of consent in the body and, in turn, the transformation of the body into an artifact— can be illustrated with a brief sequence of passages from the philosophy of volition.

In philosophic writings about the will, the enactment of freedom is often couched in terms of physical movement. Thus, the philosopher Gilbert Ryle argues that "the prime function of volitions, the task for the performance of which they were postulated, is to originate

[85]

bodily movement." Volitions, he says, are talked about as "special acts … 'in the mind' by means of which the mind gets its ideas translated into facts."[2] William James's whole chapter on the will in *The Principles of Psychology* proceeds through an analysis of physical movement: "the only ends which follow *immediately* upon our willing seem to be movements of our bodies."[3] Were I to will this room to become blue, it would require many intermediate steps to bring that blueness about; were I instead to will my arm to go up, it would immediately do so. For Augustine in *The City of God*, the will is present not only in this movement of the limbs and the complex elaborations of walking, but also in the exercise of mouth, face, ears, breath, and voice;[4] and moving still further back in time, the stoic philosopher Epictetus defines freedom as physical locomotion: "I go wherever I wish; I come from whence I wish."[5]

But if we now back up and move through this same set of writings a second time, what becomes clear is that these philosophers now imagine this process as bound up with creating. Even the brief fragment quoted from Gilbert Ryle makes audible this preoccupation: "The prime function of volitions, the task for the performance of which they were postulated, is to originate bodily movement." Ryle directs our attention away from the uses and accomplishments of volitions to a set of alternative questions about why it is we "invent" them, and what it is they "originate."[6]

The identification of the consensual with the material, and this new consensual materialism with the artifactual, is even more explicit in William James, who pictures the will as subdivided into an effort "to attend" to something, on the one hand, and an effort "to consent" to something, on the other. When he speaks of the effort "to consent" to something, one begins to hear the artifactual idiom: "So that although attention is the first and fundamental thing in volition, *express consent to the reality of what is attended to* is often an additional and quite distinct phenomenon involved." James then goes on to differentiate belief from consent: "When an idea *stings* us in a certain way, makes as it were a certain electronic connection with our self, we believe that it *is* a reality. When it stings us in another way, makes another connection with our self, we say, *let it be* reality." The grammar of "it is" and "let it be" correspond to two "peculiar attitudes of consciousness" and entails "the transition from merely considering an object as possible, to deciding and willing it to be real."[7] Consent, for James, has the seeds of the counterfactual in it; its most distinguishing feature is, in fact, its distinguishability from the stinging "it is."

This two-part structure is still more pronounced in Augustine. After he writes of how we are able to move not only our limbs but the "pliant" tissues of mouth and face, he moves inward toward the interior of the body: "In fact, even the lungs, which are the softest of all the internal organs and for that reason are protected in the cavity of the chest, are controlled by the will for the purpose of drawing breath and expelling it, and for producing and modulating the vocal sounds. In the same way as bellows serve the purpose of smiths and of organists, the lungs are obedient to the will of a man when he breathes out or breathes in, or speaks or shouts or sings." The voice is the play of the will, the regulation of breath across the interior of the softest of organs; the voice is the emergence of the artifactual out of the material base of the body, or conversely, the absorbing of the artifactual into that body. Material artifacts—bellows, the tool of smiths and organists—are inserted into the body only by way of metaphor or analogy. But it is precisely at this remarkable moment in *The City of God* that Augustine suddenly moves directly to the unambiguously literal: "Some people can even move their ears, either one at a time or both together. Others without moving the head can bring the whole scalp—all the part covered with hair—down toward the forehead and bring it back again at will. Some can swallow an incredible number of various articles and then with a slight contraction of the diaphragm, can produce, as if out of a bag, any article they please, in perfect condition."[8]

Thus, Gilbert Ryle places before us the problem of origination; William James, a stinging clarity about our own authorization of (rather than mere belief in) reality; and Augustine provides a graphic demonstration of the presence of the artifactual in the body's interior by testifying to the literal ingestion of small toys that can then be made to reappear at will. In order to illustrate this two-part structure of perception—by which the consensual is positioned in the material, and by which, in turn, the material is reconceived as artifactual—I have made use of a small sequence of passages from philosophy. But precisely that same structure emerges in many other kinds of writings, such as the medical. Richard Selzer's *Mortal Lessons: Notes on the Art of Surgery*, for example, illustrates a habit of mind visible in medical thinkers from Hippocrates through to Oliver Sacks.

Selzer describes his surgical entry into another person's body:

> There is sound, the tight click of clamps fixing teeth into severed blood vessels, the snuffle and gargle of the suction machine clearing the field of blood for the next stroke [of the knife], the litany of monosyllables with which one prays his way down and in: *clamp, sponge, suture, tie, cut*. And there is color. The green of the cloth, the white of the sponges, the red and yellow of the body. Beneath the fat lies the fascia, the tough fibrous sheet encasing the muscles. It must be sliced and the red beef of the muscles separated.[9]

A few moments later, he carries us to the deepest level.

> Deeper still. The peritoneum, pink and gleaming and membranous, bulges into the wound. It is grasped with forceps, and opened. For the first time we can see into the cavity of the abdomen. Such a primitive place. One expects ... [10]

Expects what? I expected something superlatively organic—organs, beef, eels, or something more primordial still, bacterial slime perhaps, certainly something animate, ancient, alive. Instead, this is what Richard Selzer writes:

> For the first time we can see into the cavity of the abdomen. Such a primitive place. One expects to find drawings of buffalo on the walls.[11]

The cave drawing at the center of the body is, however ancient, magnificently artifactual. So, too, is the bric-a-brac of objects and small carpenter's tools that continually turn up inside the human body in Oliver Sacks's work. Such objects can be found in medical writings from the luminous silver hydromel of Hippocrates to the present.

Elsewhere I have argued that this cultural reflex of picturing artifacts

inside the body becomes especially pronounced in the work of
Renaissance artists, philosophers, and physicians. Poets such as John
Donne and Shakespeare, visual artists such as Leonardo da Vinci, the
theologian and aesthetician Nicholas of Cusa, physicians such as
Kircher, Paré, and Daniel Sennert all continually practice the imagistic
insertion of objects into the body. The artifacts tend to be (not just any
ordinary object but) the most potent of Renaissance tools—compass,
lens, book, page, window glass. Each enters not into a diffusely or
amorphously registered body but into a highly specified bodily location.
The pair of drawing compasses, for example, so familiar from John
Donne's poem "A Valediction: Forbidding Mourning" reappears in
some of his sermons: after giving, in one of those sermons, a graphic
description of the sheer matter of the human tongue as a physical
organ, Donne suddenly inserts into the tongue an image of the drawing
compasses. Leonardo, too, has given us a drawing of the human brain,
into the center of which he has positioned a pair of drawing compasses,
as elsewhere he implicitly inserts the tool into the arms and legs of the
man whose extended limbs touch the circumference of the circle they
themselves (like a compass) seem to have brought into being. Nicolas of
Cusa, in his philosophic discourse *De Mente*, has a verbal image of a
compass residing in the brain that is close to Leonardo's. Other potent
artifacts also take up residence in the body. In "Devotions Upon
Emergent Occasions," John Donne, heavy with sickness, lies listening to
a bell in a church; soon he imagines the bell migrating into the soft,
folded robes of a priest; from there it moves into the bowels of another
sick man; then the bell applies "its gold" from that other man's bowels
to Donne's own person. This enraptured, many-page-long sequence of
events is the prelude to the climactic sentence known even to those
who have never read John Donne: "... therefore never send to know for
whom the bell tolls; it tolls for thee."[12]

    Within the Renaissance, the reflex of what I am calling "consensual
materialism" can be seen in artifacts even more varied than those
ingested by Augustine's talented swallower: lens, compass, windowpane,
beehive, spoon, stories, church bells, music. For the few pages that
remain, I want to sketch how this same reflex of consensual materialism

also becomes visible in another way. Rather than look at an *array* of interiorized objects within a solitary historical moment (the Renaissance), I want to do the opposite: to concentrate on a solitary object that seems to be fairly pervasive across an *array* of historical periods and cultures. The object is a doll, a doll either ingested into the center of the body or else made adjacent to the body as though it were one of the body's limbs. Indeed, though I have promised to speak about a solitary object, that object seems itself made up of two. It is, properly speaking, a hybrid artifact, sometimes appearing as a doll, sometimes appearing as an artificial limb. Because a doll and an artificial limb are each a species of artificial person, their conflation is perhaps not wholly surprising.

The doll-limb hybrid makes a particularly lavish and elaborate appearance in the tradition of Romantic continental writers that includes Rousseau, Kleist, Dickens, Hoffmann, Baudelaire; and it emerges perhaps most beautifully in Rilke. But it occupies a much wider chronological framework that begins with Homer and extends into current science fiction films; it also occupies a much wider cultural frame, turning up, for example, in eighteenth- and nineteenth-century Japanese theater, as well as in the Indian epics of Guatemala. Although my examples are drawn mainly from literature and medicine, the limb-doll survives in other mental worlds, such as the writings of the social contract theorists.

The limb-doll hybrid suddenly appears in Kleist's essay "On the Marionette Theater" at the moment when the extraordinarily graceful movements of the marionettes provoke Mr. C to ask, "Have you ever heard of those artificial legs made by English craftsmen for those unfortunate individuals who have lost their limbs?"[13] Explicit references to the artificial legs disappear from the essay almost as quickly as they were invoked, but their presence is sustained by the overarching frame of Kleist's essay. The tripartite structure begins with the account of the marionettes, moves at its center to a description of a statue, "The Thornpuller," then concludes with the picture of the dancing bear: thus the middle portrait—the statue forever attending to his wounded foot—embeds the image of the impaired limb into the lingering fact of the doll-like statuary.

The kinship between artful limbs and dolls frames Dickens's work, too, occurring as a major trope for the nature of imagining from his early novel *The Old Curiosity Shop* to the late novel *Our Mutual Friend*. The latter provides an epic exploration of the hybrid carried through two artisan figures, the wooden-legged literary man, Silas Wegg, and the doll's dressmaker, Jenny Wren. Although Dickens may appear to have split the hybrid—conferring the artificial limb on one person and the dolls on another—in fact the narrative repeatedly converts each into its other half: Wegg's leg becomes a doll; the dollmaker's dolls become prosthetic limbs.

The sense of Wegg's leg as a ventriloquist doll is invited by the key scene in which Boffin and Wegg make a literary contract: Wegg becomes the Literary Man who will read to Boffin long hours of the night, allowing "worlds of print to open before him." In their opening conversation, Boffin asks Wegg how he likes his leg, and contemplates the fact that its temperature is discontinuous with the rest of the body. Boffin has arrived with his own wooden limb, "carrying his knotted stick in his left arm as if it were a baby" and speaking to it, hugging it, looking down at it, while they discuss the properties of Silas Wegg's leg:

> "How did you get your wooden leg?"
> "In an accident."
> "Do you like it?"
> "Well! I haven't got to keep it warm … "
> "He hasn't," repeated the other to his knotted stick, as he gave it a hug, he hasn't got—ha! ha!—to keep it warm!"[14]

It is across this double animation of wooden limbs that their contract for a sequence of literary evenings is made, the contract that allows worlds of print to fall open before them.

Silas Wegg's comrades also have ventriloquized limbs. Boffin's doll-arm, present in this opening scene, reappears periodically, particularly if he is walking and, even more particularly, if he happens to be thinking about Wegg or Venus: he repeatedly "nurses his stick" in two arms, "glances at that companion," and "whispers" confidential remarks to it.[15] Mr. Venus, a shopkeeper of odd objects, including missing limbs

(Wegg first comes there looking for his amputated leg), has himself a ventriloquized arm with not wholly controllable "speaking properties." Venus's arm, not his voice, is his "expressive countenance"; Wegg periodically addresses the arm directly, as though it were an autonomous person.[16]

As Silas's leg and the arms of his comrades are animated dolls, so the doll's dressmaker, Jenny Wren, brings before us a story of impaired limbs. She is a child whose back and limbs are always "queer" and often in pain and who, when in motion, always has as "equipage" her little crutch-stick, her own artificial limb that, like her dolls, is an expressive agent of determination or entreaty.[17] The crutch is assigned quoted dialogue in the book distinct from the dolls' dressmaker's own sentences: separate paragraphs of words float free from Jenny Wren and are attributed to "the brisk little crutch stick." As Jenny's crutch is her animated doll, so her dolls are her prosthetic limbs: they lend her their motion. She looks at her creations one night in "a brilliantly-lighted toy-shop window" and sees "a dazzling semicircle of dolls in all the colors ... dressed ... for going to balls, for going out driving, for going out on horseback, for going out walking, for going to get married, for going to help other dolls to get married, for all the gay events of life."[18] They are her agents of surrogate motion, dolls for "going," for "going," for "going."

Dickens's two artists—his literary man and his artisan—dwell continually on the crossroads of prosthesis and puppet. Dickens locates in the doll a set of central questions about the aesthetic: he shares the bias of Baudelaire who, in the essay "A Philosophy of Toys" writes, "I have moreover retained a lasting affection and a reasoned admiration for that strange statuary art which, with its lustrous neatness, its blinding flashes of color, its violence in gesture and decision of contour, represents so well childhood ideas about beauty."[19]

Although Baudelaire, Kleist, and Dickens are located in the nineteenth century, the perceptual arc that is continually enacted—the connection of a person with a doll or puppet and the transformation of that doll into a prosthetic limb—recurs in many other periods, examples of which will be briefly summoned here.

A striking instance is the Greek god Hephaestus, artificer of many resonant cultural objects: the shield of Achilles, the bow of Philoctetes, the scepter of Agamemnon, the necklace of Harmonia, the crown of Ariadne. Though his creations are remarkable for their sheer number, we are often made to attend to any one of them in primal isolation, as though it were the only object in the world: Book 18 of the *Iliad* describes the making of Achilles' shield; the Trojan War is interrupted while we contemplate, for the duration of Sophocles' *Philoctetes*, the nature of the miraculous bow. But what we see Hephaestus make for *himself* is none of these objects—not a shield, not a bow, not a necklace, not a throne, not a scepter—but instead a pair of prosthetic dolls. Wanting crutches to support his fragile legs, Hephaestus makes two golden statues which become maidens enabling him to walk and to work.

Though Hephaestus makes giant workshops for himself at Crete and Lamos, the crutch-maidens are in effect his miniature workshop, the condition of all subsequent creation. Throughout the *Iliad*, the Homeric epithet for Hephaestus is "god of the two strong arms." He walks haltingly with "stout staff" until assisted by "handmaidens wrought of gold in the semblance of living maids." This picture is superimposed over the one Homer has given us, ten lines earlier, of Hephaestus rising from his anvil: "a huge, panting bulk, halting the while, but beneath him his slender legs moved nimbly." The entire scene of the making of the shield of Achilles—occurring in the same book in which the crutch-dolls appear—is haunted by prosthetic movement. When the mother of Achilles, "silver-footed Thetis," comes to request the shield, Hephaestus is "fashioning tripods, twenty in all, to stand around the wall of his well-built hall, and golden wheels had he set beneath the base of each that of themselves they might enter the gathering of the gods at his wish and again return to his house, a wonder to behold."[20] The penultimate ring on the shield depicts the "dancing floor": the lithe young dancers "run round with cunning feet exceeding lightly, as when a potter sitteth by his wheel that is fitted between his hands and maketh trial of it whether it will run; and now again would they run in rows toward each other."

Within the frame of Hephaestus's beautiful, animated crutch-dolls, we can see that many of the other objects he makes—tools, weapons, shields—display the inadequacy of human limbs and are themselves prostheses that magnify the power of the limbs they replace. In, for example, its spun tissue of metals (three layers for the rim, five layers for the center), the shield of Achilles is an extension of, and substitute for, the young soldier's own body. It is there to catch the blows that will otherwise fall on him.

The great artificer, Daedalus, is associated with limbs as well. Daedalus made arms and legs and attached them to the *xoana*, "the shapeless primitive statues of the gods."[21] Unlike Hephaestus, he did not make robotic or animated dolls; he did, however, as Socrates reminds us, make statues famed for looking so alive they seemed to observers to move.[22] He supplemented human arms with powerful wings in order to enable the escape from the labyrinth; and the other tools he invented, such as axe and saw, are also prosthetic transformations of human limbs.

This is true of the artificer in other cultures as well. Thus in the utensil rebellion in the Popol Vuh of Guatemala, the mud dolls and wood dolls are made by artisan gods with the elemental names Shaper and Former, the primary of whom, the Heart of Heaven, is called 1 Leg Lightning or more often 1 Leg. "Through him who is the Heart of Heaven, / 1 Leg by name. / 1 Leg Lightning is the first."[23]

This limb-doll hybrid—this combination of an artifactual person and a natural person—expresses features of art in particular and made culture in general. It is clear that Hephaestus, Daedalus, 1 Leg, Jenny Wren, Silas Wegg, Baudelaire's and Kleist's and Rilke's toy lovers are all artists (Baudelaire tells us the toy is our initiation into art; not to love beautiful bright objects, he asserts, is to be antipoetic.) But specifying artistic production matters simply because it centers our attention on the nature of production or creation in general, the making of all objects, the whole world bodied forth in the "worlds of print" in the evenings of Boffin and Wegg, the whole world of made civilization forged on the surface of Achilles' shield.

Although I have mainly here invoked literary stories, the actual artifact of the prosthetic arm is equally revealing: in fact, the starting point of my research into the doll-limb hybrid was not Homer or Dickens or the Popol Vuh but the engineering dilemmas faced by those who seek to design a prosthetic limb, especially an artificial arm, which appears to be many magnitudes of difficulty beyond the prosthetic leg.[24] The medical engineering accounts and the literary accounts of the hybrid are mutually vivifying because they often expose identical features, only a small number of which are possible to convey here. The booklets of inventions for chassis, go-carts, and other small vehicles for children with missing limbs or hurt spines almost read, in their generous ingenuity, as though they were written in the workshop of Hephaestus.

Within the medical community, both nationally and internationally, those physicians who work with children with artificial limbs repeatedly speak of the need to turn the limb into, in effect, a doll—to *animate* it in order to transform it from "an inert supplement" or "an extracorporeal structure" into a corporeal one.[25] Thus they enumerate procedures for establishing "fondness" for the object, such as having the child sleep with it;[26] and some therapists have told me that children sometimes bestow doll-like names on the limb. Because children's relations to dolls are modeled on their relation to adults in their world, it is not surprising that their relation to the artificial limb has also been found to be modeled on their mother's expression of "rapport" with them.[27]

An equivalent manifestation of the doll-limb hybrid occurs in accounts of adult limbs. The medical literature attends more elaborately to this question for children than for the many adults affected.[28] Yet even the literature about adults repeatedly notes a problem of "bonding" or "rapport" that is continuous with the problem found among children. The natural limb, when still a part of the body, is thought of as only a *part* of the body. But once severed, it is remembered as though it were a whole person: one patient expressed her relation to her limb by noting, "We were very close" (much as Franz Bieberkopf in *Berlin Alexanderplatz* holds conversations with his missing arm as if an intimate comrade; or as Silas Wegg in *Our Mutual Friend* makes repeated returns to the shop that might house his missing leg). Elaborate studies have been done by the medical community on the analogy between grieving for a lost spouse and grieving for a lost limb: both entail "pining for," disorganization around, and visual or tactile hallucinations of the person or limb missing.[29] Given the perception of the missing natural limb as a whole person, the perception of its replacement as an "artificial person" or "doll" is less surprising.

A key attribute of this hybrid is its overt and continual rehearsal of the path back and forth from inside to outside to inside once more. In the Renaissance, artifacts (compass, glass, book) are set inside the body in order to bring about consensual materialism, but they are not themselves self-conscious enactments of that mediating pathway. They simply seem to have already arrived at the body's center. The limb-doll hybrid is remarkable for the explicitness with which it traces and retraces that path. Though this feature is richly evident in many literary and medical accounts, its complexity and tone can be illustrated in a compressed space with Rilke's essay "Reflection on Dolls," and the poem so intimately attached to it, "The Fourth Elegy."

Rilke sometimes images the doll as an extension of the child's arm, so ardently connected it can almost not be unconnected, clasped fiercely in the sick child's hand. But more often he places the doll deep inside us: "Made a confidant, a confederate…initiated into the first, nameless experiences of their owners, lying about…as in the midst of empty rooms, as if all they had to do was to exploit unfeelingly the new spaciousness with all their limbs—taken into cots, dragged into the heavy folds of illnesses, present in dreams, involved in the fatalities of nights of fever such were these dolls."[30] The heavy folds of illness at first seem ambiguously the folds of bodily tissue and the folds of the blankets. But at a later moment in this beautiful essay they become unambiguously the folds of the body: "For hours together, for whole weeks we were content to lay the first downlike silk of our hearts in folds against this motionless mannequin" until our "twofold inspirations flagged, and suddenly we sat facing it, expecting something from it."[31] Deeply grafted into interior folds of body, it is continually ripped out again: "planted in the softest depth of a tenderness infinitely experimental and torn out again a hundred times, flung into a corner amongst sharp-edged broken objects, scorned, spurned, done with."[32] We take it into the interior; then violently divest ourselves of it once more, crying that it has been "unmasked as the horrible foreign body on which we had wasted our purest ardour."[33]

We continually incorporate, then repudiate, then reincorporate the artifact. Etymologically, the word "limb" is from the Latin "limbus," for border or edging. While in Rilke there is a repeated and overt locating and relocating of the object (now at the edge, now the interior, now the edge once more), the doll-limb hybrid itself, by virtue of being part limb, makes that pathway from inside to outside a *structural* fact, present throughout the succession of its transformations.

The full-length study of which this brief lecture is a part sets out to show, first, the omnipresence of the hybrid; second, the way the self-conscious movement of the limb-doll into the body and out again displays a desire for volitional positioning; third, the inextricability of contract and creation in the portrait of the hybrid; fourth, the labor of animation; and fifth, the possibility of a revolution among artifacts that comes about precisely because we lend them the heat of our own animation.

But even this brief introduction may be instructive. In our own era, any kind of inscription on the body by a cultural construct is automatically taken as a sign that something is amiss: the phenomenon tends to be treated as wholly negative. To be sure, some instances of bodily inscription *are* negative. But it cannot be the fact of inscription itself that is the problem, since this describes all of culture. If our artifacts do not act on us, there is no point in having made them. We make material artifacts in order to interiorize them: we make things so that they will in turn remake us, revising the interior of embodied consciousness. Crucial in this project of consensual materialism is the emphasis on volitional positioning so prominent in the stories about the limb-doll, the continual alternation between taking something in, then flinging it, like Rilke's dolls, across the playroom once more. Without the feature of volitional positioning, the artifact would cease to be "an artifact": it would acquire the obdurateness of, would itself become one more instance of, the material given. It is to displace the material given with the consensual that the project of cultural creation originates.

The Human Genome Project:
A Challenge in Biological Technology

Joan H. Marks

Probably no technological feat under way today has more potential to affect our lives in the twenty-first century than the Human Genome Project, an organized, collaborative effort on the part of hundreds of scientists. Still in its infancy, this initiative was launched by the U.S. Congress in 1987 with two goals: first, to develop maps for each of the twenty-three paired human chromosomes; and second, to unravel the sequence of bases that make up the DNA of each of the chromosomes. Together these efforts will determine the structure of the human genome. The genome is therefore defined as a complete set of chromosomes containing all the genes of an organism.

All genetic information is expressed in terms of a sequence of chemicals, of which there are four different types in the DNA. These four bases represent the genetic alphabet, and the sequence in which they are arranged spells out the specific information, decoded in a complex chemical process occurring in the nucleus of each cell in the body. A genetic defect is an alteration in the normal sequence of these four chemicals.

Genetic mapping is a massive undertaking which involves staking a series of structural landmarks that can be used to locate the genes underlying our physical traits, sometimes producing disease if the structure of the genes is abnormal. This sequencing will identify the precise information in each of those genes. The human genome is thought to contain three billion bases or bits of information. It is estimated that it will take fifteen to twenty years to complete the project.

Once we understand how healthy human beings function, we will be able to explain the role of genetic factors in a multitude of diseases at the chemical level. But this new technology also raises important social

[99]

questions. Recently I happened to meet Jim Watson, codiscoverer of the double helix and director of the Human Genome Project, and I asked him what was the most important message about the human genome initiative. He said without hesitation that parents will have the choice of having second children without, for example, muscular dystrophy, fragile X syndrome, or cystic fibrosis. Dr. Watson believes the strength of the human genome initiative lies in the area of prenatal diagnosis, the study of the genetic map of a fetus before birth and the option to terminate a pregnancy that presents serious, debilitating, or life-threatening conditions. Such a choice may seem troublesome, as the issue of terminating pregnancies is controversial even in such situations.

What Mapping the Human Genome Will and Will Not Accomplish

Genes tell the cells in our body how to act; when a mutation or alteration in a gene changes the information, a cell may function improperly. Identifying the genes and their mutations will provide clues as to which gene causes which negative result. When we know which gene leads to an abnormal condition, we can screen for disease before it occurs— remove a tissue at risk for cancer, treat a patient with drugs, change a diet, or maybe eventually put a corrective gene into a cell.

Upon the completion of this project, scientists will be able to develop cures for some of the four thousand known inherited disorders and the countless diseases whose origin may be due in part to genetic malfunctions. Information about a great variety of human traits will lead to benefits that include cures for certain diseases, as well as pharmacological advances.

[100] Developing the map of the human genome means obtaining a large number of "markers," which act as signposts along the length of each chromosome. These signposts are then used as reference points for locating genes related to various human traits. Thousands of genetic markers have already been identified and millions of bases of DNA have been sequenced. However, these represent less than one percent of the total amount of DNA in the whole genome. The genes for cystic fibrosis,

hemophilia, muscular dystrophy, and fragile X mental retardation have all been identified, in addition to genes that confer susceptibility to such common diseases as cancer and atherosclerosis. Using the map to identify these disease genes is the main focus of the genome project at this time.

Identifying susceptibility of predisposition does not easily lead to cure. What it may do, however, is identify people who can alter life style in order to avoid further expression of their susceptibility. Understand-ing which genes produce which deleterious effects will greatly enhance our ability to prenatally diagnose serious defects in the fetus, presenting families with the option to terminate pregnancies that will have an adverse outcome. Today, ninety-five percent of parents who receive bad news about a growing fetus opt to terminate their pregnancy in favor of a subsequent unaffected pregnancy.

Since some scientists believe much human disease is genetic in origin, the impact of this new technology is significant, indeed. To some extent it may determine who gets sick and who stays well over the course of a lifetime.

To illustrate this point, let me repeat a story I read recently. A medical geneticist received a call on a Friday morning from a neurologist colleague. The neurologist had a woman in his office with her young son who appeared to have a fatal inherited disease. In addition, the woman was pregnant! The neurologist requested urgent help from the geneticist to make a laboratory diagnosis about the son *and* the fetus. By that evening it was determined that the boy had the fatal disease, and by Monday that the fetus did not. When this story is repeated in the twenty-first century, it will probably conclude with the statement "and by Tuesday effective therapy for the once fatal disease was started in the young boy." To summarize, then, the benefits of the genome project can be viewed as threefold: prevention, cure, and manipulation or treatment.

[101]

What will the knowledge emanating from the study of the human genome *not* be able to do for us? It is important to realize that knowing the structure of a gene will not necessarily tell us how the gene functions or how we can overcome abnormal functions of the gene.

For many of the genes, though not all, a great deal of further bio-logical and clinical research will be required to fully exploit the products or potential of the genome project. But the genome project will not enable us to predict intelligence, creativity, or athletic skill, nor will it tell us who will win an Oscar, pitch a no-hitter, graduate *summa cum laude* from Harvard, or live to be one hundred.

Who Will Use the Information Generated from the Genome Initiative?

The first group to benefit from information about disease susceptibility might be employers. From a positive perspective, employers might be able to avoid placing people in work environments detrimental to health. On the other hand, this could be seen as a limit to autonomy and individual rights, limiting areas of employment.

In the educational arena, schools will know which students have a predisposition to learning disabilities, saving countless teacher hours and simultaneously enabling schools to work more effectively—and sooner—at remedial levels. Those engaged in market strategies and social planning will benefit from better information about who will get sick and when, and the types of treatments that they will need. They will pay even closer attention to individuals who are at risk for costly disabilities. In developing community resources and allocating scarce health facilities, the government will find the information emanating from the genome initiative to be useful. The courts will find the information useful for identifying those with thought disorders or criminal behavior. Accurate paternity testing now easily identifies true biological fatherhood.

[102]   Insurance companies will find it both useful and economical to know who will make premature claims. If motivated, insurance compa-nies might also focus efforts to promote wellness behavior for their high-risk clients. For example, coronary heart disease patients could be counseled to change their diet and smoking habits in exchange for reduced insurance premiums.

Problems Associated with the Human Genome Initiative

What are the dangers of this multibillion-dollar, multinational research effort? The greatest pitfall of the human genome initiative is that it is funded by the U.S. government, which is, therefore, in a strong position to control the use of the information generated by the project. Another danger is the inaccurate application of information. Without validation, some individuals could be told that they will develop disease when, in fact, they will not. Others might be given a clean bill of health when in reality they should be taking precautions from childhood. Another serious problem might occur when we have the capacity to identify people at risk for expressing an inherited disease for which there is no known cure. The anxiety raised in these cases is worrisome; some at-risk individuals, however, tell us that it is an enormous relief to finally learn the actual status of their situation after many years of anxiety. This holds true whether one learns one is or is not at risk.

Major complications can result from learning one's at-risk status, particularly for serious diseases that manifest themselves later in life. The stigma can affect one's personal and professional life. At another level, one has to ask how this new information could shape public opinion and policy in matters of health-care planning, community resources, and personal freedom.

The major worry about the human genome initiative is the accumulation of vast amounts of genetic information. This information will be difficult to control and may not always be used constructively. Insurance companies, employers, and, perhaps, other social institutions may use this information to target people with genetic susceptibility. This kind of stigmatization can have devastating effects on one's concept of self-worth, effects that can easily be transmitted to other family members. In addition, many people are concerned that our very notion of what is *normal* can be affected by our concept of what is not acceptable or different. Even in a situation where educational planning may be logical once a genetic diagnosis is learned, the effort can also be accompanied by discrimination within the educational system. Obviously, the circular effects of this kind of situation can be insidious.

In terms of life and health insurance, genetic discrimination by insurance carriers can have dire consequences beyond the obvious economic ones; people could be rendered virtually unemployable and uninsurable. Premature revelation of one's genetic status, particularly in the case of conditions that will develop in later life, cannot only alter one's own self-image but can place one at risk of discrimination from many sources.

The common perception that total understanding of the human genome will provide us with the means to manipulate genes in order to uniformly produce perfect progeny is both inaccurate and dangerous. Such experiments to manipulate genes will be flawed—we do not know how to define perfection. The idea that we can manipulate genes to produce only healthy babies is a misperception based on a faulty understanding of basic biology. One must understand that there are two types of cells in the human body: a germ-line cell that passes hereditary factors to one's progeny through the sperm and the egg; and somatic or body cells where alterations in the genes cannot be passed on to one's progeny. Altering life so that a healthy baby is produced requires the introduction of many genes into the germ line, not into the body's cells, a difference not widely understood. At present, there is a voluntary ban among scientists on the manipulation of germ-line cells. This ban should be encouraged by public and scientific quarters but must not be expanded into a ban on further elucidation of the human genome. Scientific inquiry cannot and should not be prevented from moving ahead.

What Challenge Does the Availability of Genetic Information Represent?

How could the genome experiment fail to have positive results? The promise of less disease, less human suffering, and less expense seems attractive to say the least, but to many this growing mastery over the biological world might seem not only grandiose but also against the natural order. Public discussion of the implications of genetic research is not only important but essential in light of the new knowledge already available to us. Are the issues too complicated, too technical, to be influenced by the general public? Should scientists be encouraged to make decisions about studying the human genome without guidance from nonscientists?

The social implications of many of the potential uses of information generated by the human genome initiative are complex and demand great scrutiny by the public. The quality of information must be good. The biological significance of this information must be understood *before* predictions are made. The amount of information on individuals that will be available will be unprecedented compared to what is available today, so it will be necessary for scientists to know *how* different genes interact with one another in order to allow for accurate predictions. The challenge will be to generate greater public participation than ever before to preserve citizen autonomy, and the management and understanding of a mass of information. To achieve this, the quality of health and education must improve and must be provided at an early age. Guidelines must be developed for institutions using genetic information to insure fair application. While we have the technological capacity to communicate information to millions of our citizens, can we meet the challenge of *effectively* communicating genetic information so that individuals and the public can collectively make appropriate and informed decisions?

This century has seen major advances in the prevention of disease by public health measures designed to limit the scourge of microbial disease. Death from scarlet fever is unknown in 1992. Sanitariums are forgotten mausoleums and the blood supply is now safe. But parents who once spent sleepless nights worrying about their children succumbing to whooping cough now lie awake worried about their amino results.

The knowledge of our genetic makeup can mitigate the host factors that cause disease if we can exploit our new understanding in an appropriate and positive way. At a minimum it is essential that the Human Genome Project makes an effort to educate the public about each new incremental understanding of our genome, in terms of both benefits and dangers. A greater sense of urgency with respect to an effective educational program must be developed. The potential of the Human

[105]

Genome Project can only be fully realized when the communications industry itself recognizes the economic gains—as well as the human benefits—to be earned by transmitting messages implicit in the human genome initiative. Despite the burdens of this knowledge, we can be wise consumers, healers, and advocates if we approach the implications of this powerful new information with caution.

The Dream of the Human Genome

R. C. Lewontin

The practical outcome of the belief that what we want to know about human beings is contained in the sequence of their DNA is the Human Genome Project in the United States and, in its international analogue, the Human Genome Organization (HUGO), called by one molecular biologist "the U.N. for the human genome."

These projects are, in fact, administrative and financial organizations rather than research projects in the usual sense. They have been created over the last five years in response to an active lobbying effort by scientists that include Walter Gilbert, James Watson, Charles Cantor, and Leroy Hood, aimed at capturing very large amounts of public funds and directing the flow of those funds into an immense cooperative research program.

The ultimate purpose of this program is to write down the complete ordered sequence of As, Ts, Cs, and Gs—the four nucleotides—that make up all the genes in the human genome, a string of letters that will be three billion elements long. The first laborious technique for cutting up DNA nucleotide and identifying each one in order as it is broken off was invented fifteen years ago by Allan Maxam and Walter Gilbert, but since then the process has become mechanized. DNA can now be squirted into one end of a mechanical process and out the other end will emerge a four-color computer print-out announcing "AGGACTT..." In the course of the genome project, yet more efficient mechanical schemes will be invented, and complex computer programs will be developed to catalogue, store, compare, order, retrieve, and otherwise organize and reorganize the immensely long string of letters that will emerge from the machine. The work will be a collective enterprise of very large laboratories, "Genome Centers," that are to be specially funded for this purpose.

[107]

The project is to proceed in two stages. The first is so-called "physical mapping." The entire DNA of an organism is not one long, unbroken string but, rather, is divided into a small number of units, each of which is contained in one of a set of microscopic bodies in the cell, the chromosomes. Human DNA is broken up into twenty-three different chromosomes, while fruit flies' DNA is contained in only four chromosomes. The mapping phase of the genome project will determine short stretches of DNA sequence spread out along each chromosome as positional landmarks, much as mile markers are placed along superhighways. These positional markers will be of great use in finding where in each chromosome particular genes may lie. In the second phase of the project, each laboratory will take a chromosome or a section of a chromosome and determine the complete ordered sequence of nucleotides in its DNA. It is after the second phase, when the genome project, *sensu strictu*, has ended, that the fun begins, for biological sense will have to be made, if possible, of the mind-numbing sequence of three billion As, Ts, Cs, and Gs. What will it tell us about health and disease, happiness and misery, the meaning of human existence?

The American project is run jointly by the National Institutes of Health (NIH) and the Department of Energy in a political compromise over who should have control over the hundreds of millions of dollars of public money that will be required. The project produces a glossy-paper newsletter distributed free, headed by a coat of arms showing a human body wrapped Laocoön-like in the serpent coils of DNA surrounded by the motto "Engineering, Chemistry, Biology, Physics, Mathematics." The genome project is the nexus of all sciences. My latest copy of the newsletter advertises the free loan of a twenty-three-minute video on the project, "intended for high school age and older," featuring, among others, several of the contributors to a collection of essays about the Human Genome Project, *The Code of Codes*, and a calendar of fifty "Genome Events."

Another consequence of the conviction that DNA contains the secret of human life is the appearance of a large number of popular and semipopular books touting the wonders of the Human Genome Project.[1] None of the authors of these books seems to be in any doubt about the importance of the project to determine the complete DNA sequence of a human being. "The Most Astonishing Adventure of Our Time," say Jerry E. Bishop and Michael Waldholz; "The Future of Medicine," according to Lois Wingerson; "today's most important scientific undertaking," dictating "The Choices of Modern Science," Joel Davis declares in *Mapping the Code*.

Nor are these simply the enthusiasms of journalists. The molecular biologist Christopher Wills says that "the outstanding problems in human biology...will all be illuminated in a strong and steady light by the results of this undertaking"; the great panjandrum of DNA himself, James Dewey Watson explains in his essay in the collection edited by Daniel Kevles and Leroy Hood, that he doesn't "want to miss out on learning how life works," and "a change in our philosophical understanding of ourselves" must be worth a lot of time and money.[2] Indeed, there are said to be those who have exchanged something a good deal more precious for that knowledge.

Unfortunately, it takes more than DNA to make a living organism. I once heard one of the world's leaders in molecular biology say, in the opening address of a scientific congress, that if he had a large enough computer and the complete DNA sequence of an organism, he could compute the organism, by which he meant totally describe its anatomy, physiology, and behavior. But that is wrong. Even the organism does not compute itself from its DNA. A living organism at any moment in its life is the unique consequence of a developmental history that results from the interaction of and determination by internal and external forces, what we usually think of as "environment," which are themselves partly a consequence of the activities of the organism itself as it produces and consumes the conditions of its own existence. Organisms do not find the world in which they develop. They make it. [109] Reciprocally, the internal forces are not autonomous, but act in response to the external. Part of the internal chemical machinery of a cell is only manufactured when external conditions demand it. For example, the enzyme that breaks down the sugar lactose to provide energy for bacterial growth is only manufactured by bacterial cells when they detect the presence of lactose in their environment.

Nor is "internal" identified with "genetic." Fruit flies have long hairs that serve as sensory organs, rather like a cat's whiskers. The number and placement of those hairs differ between the two sides of a fly (as they do between the left and right sides of a cat's muzzle), but not in any systematic way. Some flies have more hairs on the left, some more on the right. Moreover, the variation between sides of a fly is as great as the average variation from fly to fly. But the two sides of a fly have the same genes and have had the same environment during development. The variation between sides is a consequence of random cellular movements and chance molecular events within cells during development, so-called developmental noise. It is this same developmental noise that accounts for the fact that identical twins have different fingerprints and that the fingerprints on our left and right hands are different. A desktop computer that was as sensitive to room temperature and as noisy in its internal circuitry as a developing organism could hardly be said to compute at all.

The scientists writing about the genome project explicitly reject an absolute genetic determinism, but they seem to be writing more to acknowledge theoretical possibilities than they are writing out of conviction. If we take seriously the proposition that the internal and external codetermine the organism, we cannot really believe that the sequence of the human genome is the grail that will reveal to us what it is to be human, that it will change our philosophical view of ourselves, that it will show how life works. It is only the social scientists and social critics, such as Kevles, who comes to the genome project from his important study of the continuity of eugenics with modern medical genetics; Dorothy Nelkin, both in her book with Laurence Tancredi and in her chapter in Kevles and Hood; and, most strikingly, Evelyn Fox Keller in her contribution to *The Code of Codes,* for whom the problem of the development of the organism is central.

Nelkin, Tancredi, and Keller suggest that the importance of the Human Genome Project lies less in what it may, in fact, reveal about biology, and whether it may in the end lead to a successful therapeutic program for one or another illness, than in its validation and reinforcement of biological determinism as an explanation of all social and individual variation. The medical model that begins, for example, with a genetic explanation of the extensive and irreversible degeneration of the central nervous system characteristic of Huntington's chorea, may end with an explanation of human intelligence, of how much people drink, how intolerable they find the social condition of their lives, whom they choose as sexual partners, and whether they get sick on the job. A medical model of all human variation makes a medical model of normality, including social normality, and dictates that we preemptively or through subsequent corrective therapy bring into line anyone who deviates from that norm.

There are many human conditions that are clearly pathological and that can be said to have a unitary genetic cause. As far as is known, cystic fibrosis and Huntington's chorea occur in people carrying the relevant mutant gene irrespective of diet, occupation, social class, or education. Such disorders are rare: 1 in 2,300 births for cystic fibrosis, 1 in 3,000 for Duchenne's muscular dystrophy, 1 in 10,000 for Huntington's disease. A few other conditions occur in much higher frequency in some populations, but are generally less severe in their effects and more sensitive to environmental conditions, as for example sickle-cell anemia in West Africans and their descendants, who suffer severe effects only in conditions of physical stress. These disorders provide the model on which the program of medical genetics is built, and they provide the human interest drama on which books like *Mapping Your Genes* and *Genome* are built. In reading them, I saw again those heroes of my youth, Edward G. Robinson curing syphilis in *Dr. Ehrlich's Magic Bullet*, and Paul Muni saving children from rabies in *The Story of Louis Pasteur*.

According to the vision of the project and its disciples, we will locate on the human chromosomes all the defective genes that plague us, and then from the sequence of the DNA we will deduce the causal story of the disease and generate a therapy. Indeed, a great many defective genes have already been roughly mapped onto chromosomes and, with the use of molecular techniques, a few have been very closely located and, for even fewer, some DNA sequence information has been obtained. But causal stories are lacking and therapies do not yet exist; nor is it clear, when actual cases are considered, how therapies will flow from a knowledge of DNA sequences.

The gene whose mutant form leads to cystic fibrosis has been located, isolated, and sequenced. The protein encoded by the gene has been deduced. Unfortunately, it looks like a lot of other proteins that are a part of cell structure, so it is hard to know what to do next. The mutation leading to Tay-Sachs disease is even better understood because the enzyme specified by the gene has a quite specific and simple function, but no one has suggested a therapy. On the other hand, the gene mutation causing Huntington's disease has eluded exact location, and no biochemical or specific metabolic defect has been found for a disease that results in catastrophic degeneration of the central nervous system in every carrier of the defective gene.

A deep reason for the difficulty in devising causal information from DNA messages is that the same "words" have different meanings in different contexts and multiple functions in a given context, as in many complex languages. No word in English has more powerful implications of action than "do." "Do it now!" Yet in most of its contexts "do" as in "I do not know" is periphrastic, and has no meaning at all. While the periphrastic "do" has no *meaning*, it undoubtedly has a linguistic *function* as a place holder and spacing element in the arrangement of a sentence. Otherwise, it would not have swept into general English usage in the sixteenth century from its Midlands dialect origin, replacing everywhere the older "I know not."

So elements in the genetic messages may have meaning, or they may be periphrastic. The code sequence GTAAGT is sometimes read by the cell as an instruction to insert the amino acids *valine* and *serine* in a protein, but sometimes it signals a place where the cell machinery is to cut up and edit the message; and sometimes it may be only a spacer, like the periphrastic "do," that keeps other parts of the message an

appropriate distance from each other. Unfortunately, we do not know how the cell decides among the possible interpretations. In working out the interpretive rules, it would certainly help to have very large numbers of different gene sequences, and I sometimes suspect that the claimed significance of the genome sequencing project for human health is an elaborate cover story for an interest in the hermeneutics of biological scripture.

Of course, it can be said, as Gilbert and Watson do in their essays, that an understanding of how the DNA code works is the path by which human health will be reached.[3] If one had to depend on understanding, however, we would all be much sicker than we are. Once, when the eminent Kant scholar, Lewis Beck, was traveling in Italy with his wife, she contracted a maddening rash. The specialist they consulted said it would take him three weeks to find out what was wrong with her. After repeated insistence by the Becks that they had to leave Italy within two days, the physician threw up his hands and said, "Oh, very well, Madame. I will give up scientific principles. I will cure you today."

Certainly an understanding of human anatomy and physiology has led to a medical practice vastly more effective than it was in the eighteenth century. These advances, however, consist of greatly improved methods for examining the state of our insides, of remarkable advances in microplumbing, and of pragmatically determined ways of correcting chemical imbalances and of killing bacterial invaders. None of these depends on a deep knowledge of cellular processes or on any discoveries of molecular biology. Cancer is treated by gross physical and chemical assaults on the offending tissue. Cardiovascular disease is treated by surgery whose anatomical bases go back to the nineteenth century, by diet, and by pragmatic drug treatment. Antibiotics were originally developed without the slightest notion of how they do their work. Diabetics continue to take insulin, as they have for sixty years, despite all the research on the cellular basis of pancreatic malfunction. Of course, intimate knowledge of the living cell and of basic molecular processes may be useful eventually, and we are promised over and over that results are just around the corner. But, as Miss Adelaide so poignantly complained,

> You promise me this
> You promise me that
> You promise me everything
> under the sun.
>
> When I think of the time gone by
> I could honestly die.

Not the least of the problems of turning sequence information into causal knowledge is the existence of large amounts of "polymorphism." While the talk in most of these books is of sequencing the human genome, every human genome differs from every other. The DNA I got from my mother differs by about one-tenth of one percent, or about three million nucleotides, from the DNA I got from my father, and I differ by about that much from any other human being. The final catalogue of "the" human DNA sequence will be a mosaic of some hypothetical average person corresponding to no one. This polymorphism has several serious consequences. First, all of us carry one copy, inherited from one parent, of mutations that would result in genetic diseases if we had inherited two copies. No one is free of these, so the catalogue of the standard human genome after it is compiled will contain, unknown to its producers, some fatally misspelled sequences that code for defective proteins or no protein at all. The only way to know whether the standard sequence is, by bad luck, the code of a defective gene is to sequence the same part of the genome from many different individuals. Such polymorphism studies are not part of the Human Genome Project, and attempts to obtain money from the project for such studies have been rebuffed.

Second, even genetically "simple" diseases can be very heterogeneous in their origin. Sequencing studies of the gene that codes for a critical protein in blood-clotting has shown that hemophiliacs differ from people whose blood clots normally by any one of 208 different DNA variations, all in the same gene. These differences occur in every part of the gene, including bits that are not supposed to affect the structure of the protein.

The problem of telling a coherent causal story, and of then designing a therapy based on knowledge of the DNA sequence in such a case, is that we do not know, even in principle, all of the functions of the different nucleotides in a gene, or how the specific context in which a nucleotide appears may affect the way in which the cell machinery interprets the DNA; nor do we have any but the most rudimentary understanding of how a whole functioning organism is put together from its protein bits and pieces. Third, because there is no single, standard, "normal" DNA sequence that we all share, observed sequence differences between sick and well people cannot, in themselves, reveal the genetic cause of a disorder. At the least, we would need the sequences of many sick and many well people to look for common differences between sick and well. But if many diseases are like hemophilia, common differences will not be found and we will remain mystified.

The failure to turn knowledge into therapeutic power does not discourage the advocates of the Human Genome Project because their vision of therapy includes *gene* therapy. By techniques that are already available and need only technological development, it is possible to implant specific genes containing the correct sequence into individuals who carry a mutated sequence, and to induce the cell machinery of the recipient to use the implanted genes as its source of information. Indeed, the first case of human gene therapy for an immune disease—the treatment of a child who suffered from a rare disorder of the immune system—has already been announced and seems to have been a success. The supporters of the genome project agree that knowing the sequence of all human genes will make it possible to identify and isolate the DNA sequences for large numbers of human defects which could then be corrected by gene therapy. In this view, what is now an ad hoc attack on individual disorders can be turned into a routine therapeutic technique, treating every physical and psychic dislocation, since everything significant about human beings is specified by their genes.

However, gene implantation may affect not only the cells of our temporary bodies, our *somatic* cells, but the bodies of future generations through accidental changes in the *germ* cells of our reproductive organs. Even if it were our intention only to provide properly functioning genes to the immediate body of the sufferer, some of the implanted DNA would be wreaked on our descendants to the remotest time. So David Suzuki and Peter Knudtson make it one of their principles of "genethics" (they have self-consciously created ten of them) that "while genetic manipulation of human somatic cells may lie in the realm of personal choice, tinkering with human germ cells does not. Germ-cell therapy, without the consent of all members of society, ought to be explicitly forbidden."[4]

Their argument against gene therapy is a purely prudential one, resting on the imprecision of the technique and the possibility that a "bad" gene today might turn out to be useful someday. This seems a slim base for one of the Ten Commandments of biology, for, after all, the techniques may get a lot better and mistakes can always be corrected by another round of gene therapy. The vision of power offered to us by gene therapists makes gene transfer seem rather less permanent than a silicone implant or a tummy tuck. The bits of ethics in *Genethics* is, like a Unitarian sermon, nothing that any decent person could quarrel with. Most of the "genetic principles" turn out to be, in fact, prudent advice about why we should not screw around with our genes or those of other species. While most of their arguments are sketchy, Suzuki and Knudtson are the only authors among those under review who take seriously the problems presented by genetic diversity among individuals, and who attempt to give the reader enough understanding of the principles of population genetics to think about these problems.

Most death, disease, and suffering in rich countries do not arise from muscular dystrophy and Huntington's chorea, and, of course, the majority of the world's population is suffering from one consequence or another of malnutrition and overwork. For Americans, it is heart disease, cancer, and stroke that are the major killers, accounting for seventy percent of deaths, and about sixty million people suffer from chronic cardiovascular disease. Psychiatric suffering is harder to estimate, but before the psychiatric hospitals were emptied in the 1960s, there were 750,000 psychiatric inpatients. It is now generally accepted that some fraction of cancers arise on a background of genetic predisposition. That is, there are a number of genes known, the so-called oncogenes, that have information about normal cell division. Mutations in these genes result (in an unknown way) in making cell division less stable and more likely to occur at a pathologically high rate. Although a number of such genes have been located, their total number and the proportion of all cancers influenced by them is unknown.

In no sense of simple causation are mutations in these genes *the* cause of cancer, although they may be one of many predisposing conditions. Although a mutation leading to extremely elevated cholesterol levels is known, the great mass of cardiovascular disease has utterly defied genetic analysis. Even diabetes, which has long been known to run in families, has never been tied to genes and there is no better evidence for a genetic predisposition to it in 1992 than there was in 1952 when serious genetic studies began. No week passes without the announcement in the press of a "possible" genetic cause of some human ill which upon investigation "may eventually lead to a cure." No literate public is unassailed by the claims. The *Morgunbladid* of Reykjavik asks its readers rhetorically, "*Med allt i genumon?*" ("Is It All in the Genes?") in a Sunday supplement.

The rage for genes reminds us of Tulipomania and the South Sea Bubble in McKay's *Great Popular Delusions of the Madness of Crowds*. Claims for the definitive location of a gene for schizophrenia and manic depressive syndrome using DNA markers have been followed repeatedly by retraction of the claims and contrary claims as a few more members of a family tree have been observed, or a different set of families examined. In one notorious case, a claimed gene for manic depression, for which there was strong statistical evidence, was nowhere to be found when two members of the same family group developed symptoms. The original claim and its retraction both were published in the international journal *Nature*, causing David Baltimore to cry out at a scientific meeting, "Setting myself up as an average reader of *Nature*, what am I to believe?" Nothing.

Some of the wonder-rabbis and their disciples see even beyond the major causes of death and disease. They have an image of social peace and order emerging from the DNA databank at the NIH. The editor of the most prestigious general American scientific journal, *Science*, an energetic publicist for large DNA sequencing projects, in special issues of his journal filled with full-page multicolored advertisements from biotechnology equipment manufacturers, has visions of genes for alcoholism, unemployment, domestic and social violence, and drug addiction. What we had previously imagined to be messy moral, political, and economic issues turns out, after all, to be simply a matter of an occasional nucleotide substitution. While the notion that the war on drugs will be won by genetic engineering belongs to Cloud Cuckoo

[117]

Land, it is a manifestation of a serious ideology that is continuous with the eugenics of an earlier time.

Daniel Kevles has quite persuasively argued in his earlier book on eugenics that classical eugenics became transformed from a social program of general population improvement into a family of programs providing genetic knowledge to individuals facing reproductive decisions. But the ideology of biological determinism on which eugenics was based has persisted and, as is made clear in Kevles's excellent short history of the genome project in *The Code of Codes*, eugenics in the social sense has been revivified. This has been in part a consequence of the mere existence of the genome project, with its accompanying public relations, and the heavy public expenditure it will require. These alone validate its determinist Weltanschauung. The publishers declare the glory of DNA and the media showeth forth its handiwork.

The nine books I have mentioned here are only a sample of what has been and what is to come. The cost of sequencing the human genome is estimated optimistically at $300 million a year (ten cents a nucleotide for the three billion nucleotides of the entire genome), but if development costs are included it surely cannot be less than half a billion in current dollars. In fact, the managers of the project are hoping for a budget of $200 million a year for fifteen years. Moreover, the genome project *sensu strictu* is only the beginning of wisdom. Yet more hundreds of millions must be spent on chasing down the elusive differences in DNA for each specific genetic disease, of which some three thousand are now known, and some considerable fraction of that money will stick to entrepreneurial molecular geneticists. None of our authors has the bad taste to mention that many molecular geneticists of repute, including several of the essayists in *The Code of Codes*, are founders, directors, officers, and stockholders in commercial biotechnology firms, including the manufacturers of the supplies and equipment used in sequencing research. Not all authors have Norman Mailer's openness when they write advertisements for themselves.

It has been clear since the first discoveries in molecular biology that "genetic engineering," the creation to order of genetically altered organisms, has an immense possibility for producing private profit. If the genes that allow clover plants to manufacture their own fertilizer out of the nitrogen in the air could be transferred to maize or wheat, farmers would save great sums and the producers of the engineered seed would make a great deal of money. Genetically engineered bacteria grown in large fermenting vats can be made into living factories to produce rare and costly molecules for the treatment of viral diseases and cancer. A bacterium has already been produced that will eat raw petroleum, making oil spills biodegradable. As a consequence of these possibilities, molecular biologists have become entrepreneurs. Many have founded biotechnology firms funded by venture capitalists. Some have become very rich when a successful public offering of their stock has made them suddenly the holders of a lot of valuable paper. Others find themselves with large blocks of stock in international pharmaceutical companies who have bought out the biologist's mom-and-pop enterprise and acquired their expertise in the bargain.

No prominent molecular biologist of my acquaintance is without a financial stake in the biotechnology business. As a result, serious conflicts of interest have emerged in universities and in government service. In some cases graduate students working under entrepreneurial professors are restricted in their scientific interchanges, in case they might give away potential trade secrets. Research biologists have attempted, sometimes with success, to get special dispensations of space and other resources from their universities in exchange for a piece of the action. Biotechnology joins basketball as an important source of educational cash.

Public policy, too, reflects private interest. James Dewey Watson resigned in April as head of the NIH Human Genome Office as a result of pressure put on him by Bernardine Healey, director of the NIH. The immediate form of this pressure was an investigation by Healey of the financial holdings of Watson or his immediate family in various biotechnology firms. But nobody in the molecular biological community believes in the seriousness of such an investigation, because everyone including Dr. Healey knows that there are no financially disinterested candidates for Watson's job. What is really at issue is a disagreement about patenting the human genome. Patent law prohibits the patenting of anything that is "natural," so, for example, if a rare plant were discovered in the Amazon whose leaves could cure cancer, no one could patent it. But, it is argued, isolated genes are not natural, even though the organism from which they are taken may be. If human DNA sequences are to be the basis of future therapy, then the exclusive ownership of such DNA sequences would be money in the bank.

Dr. Healey wants the NIH to patent the human genome to prevent private entrepreneurs, and especially foreign capital, from controlling what has been created with American public funding. Watson, whose family is reported to have a financial stake in the British pharmaceutical firm Glaxo, has characterized Healey's plan as sheer "lunacy," on the grounds that it will slow down the acquisition of sequence information. (Watson has denied any conflict of interest.) Sir Walter Bodmer, the director of the Imperial Cancer Research Fund and a major figure in the European genome organization, spoke the truth that we all know lies behind the hype of the Human Genome Project when he told the *Wall Street Journal* that "the issue [of ownership] is at the heart of everything we do."

The study of DNA is an industry with high visibility, a claim on the public purse, the legitimacy of a science, and the appeal that it will alleviate individual and social suffering. So its basic ontological claim, of the dominance of the Master Molecule over the body physical and the body politic, becomes part of general consciousness. Evelyn Fox Keller's chapter in *The Code of Codes* brilliantly traces the percolation of this consciousness through the strata of the state, the universities, and the media, producing an unquestioned consensus that the model of cystic fibrosis is a model of the world. Daniel Koshland, the editor of *Science*, when asked why the Human Genome Project funds should not be given instead to the homeless, answered, "What these people don't realize is that the homeless are impaired.... Indeed, no group will benefit more from the application of human genetics."[5]

Beyond the building of a determinist ideology, the concentration of knowledge about DNA has direct practical, social, and political consequences, what Dorothy Nelkin and Laurence Tancredi call "The Social Power of Biological Information." Intellectuals in their self-flattering wish-fulfillment say that knowledge is power, but the truth is that knowledge further empowers only those who have or can acquire the power to use it. My possession of a Ph.D. in nuclear engineering and the complete plans of a nuclear power station will not reduce my electric bill by a penny. So with the information contained in DNA, there is no instance where knowledge of one's genes does not further concentrate the existing relations of power between individuals and between the individual and institutions.

When a woman is told that the fetus she is carrying has a fifty percent chance of contracting cystic fibrosis, or for that matter that it will be a girl although her husband desperately wants a boy, she does not gain additional power just by having that knowledge, but is only forced by it to decide and to act within the confines of her relation to the state and her family. Will her husband agree to or demand an abortion, will the state pay for it, will her doctor perform it? The slogan "a woman's right to choose" is a slogan about conflicting relations of power, as Ruth Schwartz Cowan makes clear in her essay "Genetic Technology and Reproductive Choice: An Ethics for Autonomy" in *The Code of Codes*.[6]

Increasingly, knowledge about the genome is becoming an element in the relation between individuals and institutions, generally adding to the power of institutions over individuals. The relations of individuals to the providers of health care, to schools, to the courts, and to employers are all affected by knowledge, or the demand for knowledge, about the state of one's DNA. In the essays by both Henry Greeley and Dorothy Nelkin in *The Code of Codes*, and in much greater detail and extension in *Dangerous Diagnostics*, the struggle over biological information is revealed. The demand by employers for diagnostic information about the DNA of prospective employees serves the firm in two ways. First, as providers of health insurance, either directly or through their payment of premiums to insurance companies, employers reduce their wage bill by hiring only workers with the best health prognoses. Second, if there are workplace hazards to which employees may be in different degrees sensitive, the employer may refuse to employ those whom it judges to be sensitive. Not only does such employment exclusion reduce the

[121]

potential costs of health insurance, but it shifts the responsibility of providing a safe and healthy workplace from the employer to the worker. It becomes the worker's responsibility to look for work that is not threatening. After all, the employer is helping the workers by providing a free test of susceptibilities and so allowing them to make more informed choices of the work they would like to do. Whether other work is available at all, or worse paid, or more dangerous in other ways, or only in a distant place, or extremely unpleasant and debilitating is simply part of the conditions of the labor market. So Koshland is right after all: unemployment and homelessness do indeed reside in the genes.

Biological information has also become critical in the relation between individuals and the state, for DNA has the power to put a tongue in every wound. Criminal prosecutors have long hoped for a way to link accused persons to the scene of a crime when there are no fingerprints. By using DNA from a murder victim and comparing it with DNA from dried blood found on the person or property of the accused, or by comparing the accused's DNA with DNA from skin scrapings under the fingernails of a rape victim, prosecutors attempt to link criminal and crime. Because of the polymorphism of DNA from individual to individual, a definitive identification is, in principle, possible. But, in practice, only a bit of DNA can be used for identification so there is some chance that the accused will match the DNA from the crime scene even though someone else is in fact guilty.

Moreover, the methods used are prone to error, and false matches (as well as false exclusions) can occur. For example, the FBI characterized the DNA of a sample of 225 FBI agents and then, on a retest of the same agents, found a large number of mismatches. Matching is almost always done at the request of the prosecutor, because tests are expensive and most defendants in assault cases are represented by a public defender or court-appointed lawyer. The companies who do the testing have a vested commercial interest in providing matches, and the FBI, which also does some testing, is an interested party.

Because different ethnic groups differ in the frequency of the various DNA patterns, there is also the problem of the appropriate reference group to whom the defendant is to be compared. The identity of that reference group depends in complex ways on the circumstances of the case. If a woman who is assaulted lives in Harlem near the borderline between black, Hispanic, and white neighborhoods at 110th Street, which of these populations or combination of them is appropriate for calculating the chance that a "random" person would match the DNA found at the scene of the crime? A paradigm case was tried last year in Franklin County, Vermont. DNA from blood stains found at the scene of a lethal assault matched the DNA of an accused man. The prosecution compared the pattern with population samples of various racial groups, and claimed that the chance that a random person other than the accused would have such a pattern was astronomically low.

Franklin County, however, has the highest concentration of Abenaki Indians and Indian/European admixture of any county in the state. The Abenaki and Abenaki/French Canadian population are a chronically poor and underemployed sector in rural Franklin County and across the border in the St. Jacques River region of Canada, where they have been since the Western Abenaki were resettled in the eighteenth century. The victim, like the accused, was half Abenaki, half French Canadian and was assaulted where she lived, in a trailer park, about one-third of whose residents are Abenaki ancestry. It is a fair presumption that a large fraction of the victim's circle of acquaintances came from the Indian population. No information exists on the frequency of DNA patterns among Abenaki and Iroquois, and on this basis the judge excluded the DNA evidence. But the state could easily argue that a trailer park is open to access from any passersby, and that the general population of Vermont is the appropriate base of comparison. Rather than objective science, we are left with intuitive arguments about the patterns of people's everyday lives.

The dream of the prosecutor, to be able to say, "Ladies and gentlemen of the jury, the chance that someone other than the defendant could be the criminal is 1 in 3,426,237," has very shaky support. When biologists have called attention to the weaknesses of the method in court or in scientific publications, they have been the objects of considerable pressure. One author was called twice by an agent of the Justice Department, in what the scientist describes as intimidating attempts to have him withdraw a paper in the press.[7] Another was asked questions about his visa by an FBI agent attorney when he testified, a third was asked by a prosecuting attorney how he would like to spend the night in jail, and a fourth received a fax demand from a federal prosecutor requiring him to produce peer reviews of a journal article he had submitted to the *American Journal of Human Genetics*, fifteen minutes before a fax from the editor of the journal informed the author of the existence of the reviews and their contents. Only one of the authors discussed here, Christopher Wills, discusses the forensic use of DNA, and he has been a prosecution witness himself. He is dismissive of the problems and seems to share with prosecutors the view that the nature of the evidence is less important than the conviction of the guilty.

Both prosecutors and defense forces have produced expert witnesses of considerable prestige to support or question the use of DNA profiles as a forensic tool. If professors from Harvard disagree with professors from Yale (as in this case), what is a judge to do? Under one legal precedent, the so-called *Frye* rule,[8] such a disagreement is cause for barring the evidence, which "must be sufficiently established to have gained general acceptance in the particular field in which it belongs." But all jurisdictions do not follow *Frye*, and what *is* "general acceptance," anyway? In response to mounting pressure from the courts and the Department of Justice, the National Research Council (NRC) was asked to form a committee on DNA technology in forensic science, to produce a definitive report and recommendations. They have now done so, adding greatly to the general confusion.[9]

Two days before the public release of the report, the *New York Times* carried a front-page article by one of its most experienced and sophisticated science reporters, announcing that the NRC committee had recommended that DNA evidence be barred from the courts. This was greeted by a roar of protest from the committee, whose chairman, Victor McKusick of Johns Hopkins University, held a press conference the next morning to announce that the report, in fact, approved of the forensic use of DNA substantially as it was now practiced. The *Times*, acknowledging an "error," backed off a bit, but not much, quoting various experts who agreed with the original interpretation. A member of the committee was quoted as saying he had read the report "fifty times" but hadn't really intended to make the criticisms as strong as they actually appeared in the text.

One seems to have hardly any other choice but to read the report for oneself. As might be expected, the report says in effect, "none of the above," but in substance it gives prosecutors a pretty tough row to hoe. Nowhere does the report give wholehearted support to DNA evidence as currently used. The closest it comes is to state: "The current laboratory procedure for detecting DNA variation…is *fundamentally* sound [emphasis added]…It is now clear that DNA typing methods are a most powerful adjunct to forensic science for personal identification and have immense benefit to the public." Further on it asserts, "DNA typing is capable, *in principle*, of an extremely low inherent rate of false results" (emphasis added). Unfortunately for the courts looking for assurances, these statements are immediately preceded by the following: "The committee recognizes that standardization of practices in forensic laboratories in general is more problematic than in other laboratory settings; stated succinctly, forensic scientists have little or no control over the nature, condition, form, or amount of sample with which they must work." Not exactly the ringing endorsement suggested by Professor McKusick's press conference. On the other hand, there are no statements calling for the outright barring of DNA evidence. There are, however, numerous recommendations that, if taken seriously, would lead any moderately businesslike defense attorney to file an immediate appeal of any case lost on DNA evidence. On the issue of laboratory reliability the report says: "Each forensic-science laboratory engaged in DNA typing must have a formal, detailed quality-assurance and quality-control program to monitor work," and also: "Quality-assurance programs in individual laboratories alone are insufficient to ensure high standards. External mechanisms are needed…. Courts should require that laboratories providing DNA typing evidence have proper accreditation for each DNA typing method used."

The committee then discusses mechanisms of quality control and accreditation in greater detail. Since no laboratory currently meets those requirements and no accreditation agency now exists, it is hard to see how the committee's report can be read as an endorsement of the current practice of presenting evidence. On the critical issue of population comparisons the committee actually uses legal language sufficient to bar any of the one-in-a-million claims that prosecutors have relied on to dazzle juries: "Because it is impossible or impractical to draw a large enough population to test directly calculated frequencies of any particular profile much below 1 in 1,000, there is not a sufficient body of empirical data on which to base a claim that such frequency calculations are reliable or valid." "Reliable" and "valid" are terms of art here, and Judge Jack Weinstein, who was a member of the committee, certainly knew that. This sentence should be copied in large letters and hung framed on the wall of every public defender in the United States. On balance, the *New York Times* had it right the first time. Whether by ineptitude or design, the NRC Committee has produced a document rather more resistant to spin than some may have hoped.

In order to understand the committee's report, one must understand the committee and its sponsoring body. The National Academy of Sciences is a self-perpetuating honorary society of prestigious American scientists, founded during the Civil War by Lincoln to give expert advice on technical matters. During World War I, Woodrow Wilson added the National Research Council as the operating arm of the Academy, which could not produce from its own ranks of eminent ancients enough technical competence to deal with the growing complexities of the government's scientific problems. Any arm of the state can commission an NRC study; the present one was paid for by the FBI, the NIH Human Genome Center, the National Institute of Justice, the National Science Foundation, and two nonfederal sources, the Sloan Foundation and the State Justice Institute.

Membership in study committees almost inevitably includes divergent prejudices and conflicts of interest. The Forensic DNA Committee included people who had testified on both sides of the issue in trials and at least two members had clear financial conflicts of interest. One was forced to resign near the end of the committee's deliberations when the full extent of his conflicts was revealed. A preliminary version of the report, much less tolerant of DNA profile methods, was leaked to the

FBI by two members of the committee, and the Bureau made strenuous representations to the committee to get them to soften the offending sections. Because science is supposed to find objective truths that are clear to those with expertise, NRC findings do not usually contain majority and minority reports, and, of course, in the present case a lack of unanimity would be the equivalent of a negative verdict. So we may expect reports to contain contradictory compromises among contending interests, and public pronouncements about a report may be in contradiction to its effective content. *DNA Technology in Forensic Science* in its formation and content is a gold mine for the serious student of political science and scientific politics.

There is no aspect of our lives, it seems, that is not within the territory claimed by the power of DNA. In 1924 William Bailey published in the *Washington Post* an article about "radithor," a radioactive water of his own preparation, under the headline, SCIENCE TO CURE ALL THE LIVING DEAD, subtitled What a Famous Savant Has to Say about the New Plan to Close Up the Insane Asylums, Wipe Out Illiteracy, and Make Over the Morons by His Method of Gland Control.[10] Nothing was more up to date in the 1920s than a combination of radioactivity and glands. Famous savants, it seems, still have access to the press in their efforts to sell us, at a considerable profit, the latest concoction.

AIDS, Identity, and the Politics of Gender[1]

Paula A. Treichler

This essay begins with a joke:

> Joe is a regular at his neighborhood bar. One night he tells his
> buddies he's going to have sex-change surgery. "I just feel
> there's a woman inside me," he says, "and I'm going to let her
> out." A few months later Joe—or rather Jane—shows up at the
> bar and introduces herself to her old buddies. Once they're
> over their amazement, they greet her warmly, buy a pitcher,
> and start asking her about the surgery.
>
> "What hurt the most?" they ask. "Was it when they cut
> your penis?"
> "No," says Jane, "that wasn't what hurt the most."
> "Was it when they cut your balls?"
> "No, that wasn't what hurt the most."
> "So, what was it that hurt the most?"
> "What hurt the most," says Jane, "was when they cut
> my salary."

When I shared this with a medical school colleague, he got huffy:
"That's not a joke," he said, "that's feminism!" Indeed; it's feminism that
turns on an important paradox of identity: When does "Joe" become
"Jane"? And even *does* Joe become Jane?

    If someone is called a "woman," what criterion is being treated as
essential? Are there specific characteristics, qualities, behaviors that
count as female? Feature X but not feature Y? X chromosome but not
Y? Recent reports on the gender testing of Olympic athletes ("chick
tests") assure us that these are all unreliable criteria. "For all its dazzling

[129]

discoveries about the genes that guide a human embryo along its path to maleness or femaleness," wrote Gina Kolata in the *New York Times*, "science, it appears, cannot provide a simple answer."[2] How about having a vagina, then? Or having a "real" vagina as opposed to a constructed one? One feminist publishing house refused to allow a feminist researcher to use the pronoun "she" in reference to a male-to-female transsexual. Is birth sex forever? What if someone with a penis wears a dress, uses the ladies room, and is universally taken as female? Are these criteria independent of each other, or is one fundamental? When Joe changes, does something "in" him stay the same? Note that Joe's pals didn't treat Jane as a stranger, nor as a single woman in a bar, but as someone familiar, a drinking buddy. Yet when did Joe become Jane? When does "he" become "she"? Does it all come down to the moment "she" picks up her paycheck?

Questions of gender and identity also emerge in narratives of the AIDS epidemic. Indeed, as a small sample of media headlines reveal, familiar identities were ready and waiting for women from the beginning: the loyal companion who stands by her man ("AIDS VICTIM TO WED IN ST. PATRICK'S CATHEDRAL"); the scheming carrier who deliberately infects her male victim ("WIFE MURDERS HUBBY WITH AIDS COCKTAIL"); the protector of morals, whose draconian proposals offer the illusion of control; the Madonna ("BLESSED VIRGIN CURED MY AIDS, MAN SAYS"); the whore ("CINCY PROSTITUTES FEARED SPREADING AIDS"); the innocent victim ("Her sickness would have been easier to accept if she'd been a slut or a drug user," said Kimberly Bergalis's father, "But she had done everything right"); the transparent vessel ("AIDS CARRIER'S BABY NOT INFECTED, HOSPITAL STATES"); and above all, the loving mom, wife, or caretaker whose presence serves to humanize and desexualize the infected (gay) man, a role she shares in photographs with stuffed animals and pets.

An effective response to an epidemic (as to any widespread cultural crisis) depends on the existence of identities for whom that epidemic is meaningful and stories that take up those identities and give them life. Though plentiful, the identities AIDS scripts for women have rarely worked to expand women's own awareness of the epidemic or to further social change. Rather, facts, information campaigns, even stories of sickness have served, sometimes inadvertently, to discourage women from engaging seriously with the epidemic. Women lack access not only to clear information, in other words, but also to the subject positions, narratives, and identities that could make sense of that information and act on it. And if we, ourselves, cannot make sense of AIDS, articulate its influence upon our lives, and shape interventions that embody our interests and perspectives, history holds out little promise that anyone else will do it for us. Instead, the direction of the epidemic will be influenced by the prevailing stereotypes and ongoing confusions about identity that still, after more than a decade of documented cases among women, encourage women to feel immune to a "gay male disease."

Perhaps we should not be surprised that the crisis of women and AIDS has been systematically neglected. As Clay Stephens has written, "Most of the problems are not new; they are simply viewed through another set of distorted lenses. AIDS is a paradigm for the condition of women within our society."[3] But it is now imperative that we look closely at AIDS commentary on gender and end that neglect. Without intervention, it is hard to be optimistic about women and AIDS in the United States. The current state of affairs is that we lack knowledge, social policy, and cultural consensus—in part because we lack conceptual coherence about the role of gender in HIV transmission and about the epidemic's impact on women, families, and society at large. Women are studied in the scholarly literature on AIDS as an index to something else: total numbers infected, extent of heterosexual spread, impact on childbearing and caretaking, clues to patterns of disease transmission, or proof that women of the 1990s are not what they seem. In the popular literature, AIDS primarily provides a new occasion for recycling old narratives out of women's historical role in epidemics and disease.

I.

In January 1988, for example, physician Robert E. Gould published an article on women and AIDS in *Cosmopolitan* magazine. Seeking to reassure *Cosmo's* several million women readers about AIDS, Dr. Gould wrote that "there is almost no danger of contracting AIDS through *ordinary sexual intercourse*" (emphasis in original)—a term that means, in his words, "penile penetration of a well-lubricated vagina—penetration that is not rough and does not cause lacerations."[4] When Gould says AIDS, he means HIV, the human immunodeficiency virus. He also means *unprotected* heterosexual penile-vaginal intercourse *with an infected man*. But the article's real problem is that Gould, a psychiatrist with no special expertise in AIDS research or treatment, argues that a "healthy vagina" is protection enough against the virus. If it were not, he reasons, the prevalence of AIDS in the U.S. heterosexual population would by now be extensive.

In evaluating such gendered accounts of the physical body, we need to ask what the stakes are. Implicit in claims of fixed biological difference is not only a conviction that AIDS is uniquely homosexual but that it represents a boundary transgression, a violation of natural difference. Gould's basic claim, after all, depends on his equation of "ordinary sexual intercourse" with what he considers healthy and natural. In this he closely echoes science writer John Langone, whose influential 1985 article in *Discover* distinguished the "vulnerable rectum" from the "rugged vagina"; the vagina, he argued, "designed to withstand the trauma of intercourse as well as childbirth," is too tough for the virus to penetrate.[5] Thus gendering and naturalizing the body's vulnerability to viral penetration, Langone can conclude that AIDS "is—and is likely to remain—the fatal price one can pay for anal intercourse." But in Langone's 1988 book, *AIDS: The Facts*, though he still argued that "a woman's genital anatomy seems to be in her favor insofar as AIDS is concerned," he now emphasized that the "vagina may be rugged, but it's not all that rugged" and identified a number of conditions and "special circumstances" that could weaken a woman's "vaginal armor" and make her vulnerable to HIV, including menstruation, vaginal bruises, "other diseases and conditions," and aging![6]

Fixed biological genderings of the AIDS body appear in scientific and medical journals as well as in popular magazines like *Cosmopolitan*. One common hypothesis, for example, was that a significant quantity of virus was required to produce HIV infection, a quantity transmissable only by a specialized projectile mechanism—a penis, for example, or a syringe. Women, lacking such privileged instruments of contagion, could not provide the requisite viral jolt. When research in the mid 1980s attributed HIV infection among U.S. servicemen in Germany to contact with female prostitutes, a hot debate ensued in the *Journal of the American Medical Association* and elsewhere over whether it is *possible* for women to be infected with HIV and, especially, for women in turn to transmit the virus to men. The "medical" argument amounted to something like this: Men get it, women don't, so women *can't*. Pretty sophisticated stuff. How, then, to explain the servicemen's HIV infection? One version put it this way: Man A, already infected through an accepted mode of transmission (homosexuality, hemophilia, drug use, and so on) ejaculates into the prostitute and—because (it is taken for granted) she "performs only perfunctory cleaning"—his semen remains in her vagina, where it infects Man B during sexual intercourse. This account rejects female-to-male transmission in favor of "quasi-homosexual transmission." Even so, the projectile penis must double as a kind of proboscis that's capable of sucking out the virus from the contaminated vaginal vessel.[7]

While it is plausible to suppose that physiological factors affect the probability of HIV transmission, the statistics of the epidemic do not support claims that women's genital anatomy, without help from condoms or spermicides, constitutes "vaginal armor" against viral penetration. As early as December 1981, women accounted for more than three percent of total reported AIDS cases in the United States, and the incidence has shown a steady, gradual increase; now nearly fifteen years a decade into the epidemic, women account for more than twelve percent of the 356,275 total cases.[8] New AIDS cases in women are now growing faster than those in men, and more than half of the total cases have been reported in the last two years; current estimates are that many additional women are infected with HIV. Women of color are disproportionately affected. Some women with AIDS identify themselves as lesbian or bisexual. Though the U.S. male-female ratio is conventionally given as seven to one—seven men with AIDS to one woman—this averaged figure is misleading: such a ratio depends on the composition of the particular population being measured and can therefore vary according to exposure category, sex and sexual orientation, geography, ethnic and social subculture, class, and so on; in some communities, men with AIDS may out-number women by twenty or thirty to one; elsewhere the ratio may approach one to one. One of the things this difference reflects is the frequency and regularity with which infected men in a given community have unprotected sexual contact with women. It is therefore more accurate to think of *the* AIDS epidemic in the U.S. as a series of intersecting local epidemics, each with its own dynamic.

Statistical accounts, of course, provide information about aggregates and probabilities, whether for individuals or populations, risk groups or modes of exposure. What they don't do is convey absolute certainty about individual risk, encourage personal self-identification, or describe lived experience. They do not capture the fluidity, transgressive potential, and unpredictability that may characterize the behavior of individuals. Moreover, numerous features of AIDS surveillance reporting thwart our attempts to decipher what even the aggregate figures mean. Women are often not clearly identified *as women* in statistical reports, so data about them must be deduced indirectly from data in other categories. In the familiar pie charts that show percentages of AIDS cases represented by different groups, women were for many years invisible within official categories called "Undetermined mode of exposure," "No identified risk," or (ironically) "Other."

Yet even today, when AIDS among women is widely and officially acknowledged, women are *still* invisible. A 1991 pie chart in *Newsweek* was headed "Who Has AIDS?"[9] The total U.S. cases to date at that point were broken down into the following categories:

homosexual/bisexual males
heterosexual intravenous (IV) drug users
male homosexual/bisexual IV drug users
heterosexuals
transfusions
hemophiliacs, and
undetermined.

In addition to mixing "risk groups" with "modes of exposure" (thereby confusing identity with behavior, "who you are" with "what you do"), this list contains few clues as to which of these categories might be gendered female. There are other problems with the Centers for Disease Control's official classification of the "risk groups" to which adult and adolescent females with AIDS are assigned:

IV drug users
sexual partners of IV drug users
sexual partners of homosexual or bisexual men
transfusion recipients
sexual partners of persons with hemophilia
heterosexuals born in Pattern II countries (countries
      where heterosexual transmission is endemic), and
      mothers of pediatric AIDS patients

Even though explicitly gendered female, the CDC list does not readily translate into recognizable identities or groups. It also reinforces the longstanding view that women's identities are not autonomous but determined by the significant others to whom they are attached. And it confers a reality on categories that may have no coherence except in the researchers' minds while granting "real," lived identities no official recognition. For example:

• Lesbians are excluded from surveillance reporting and research; lesbians and bisexual women are not counted as a population at risk nor is women's same-sex behavior listed as a potential mode of HIV transmission. Lesbians' generally low incidence of sexually transmitted diseases is used to justify their exclusion, but Clay Stephens has suggested that researchers' lack of interest in lesbians also reflects a common view that sex between women is too infrequent, gentle, or boring to transmit a virus.

• Though prostitutes are neither an official "risk category" in AIDS surveillance nor are believed to constitute a significant source of sexual transmission of HIV, they continue to be scapegoated whenever "women" are mentioned. At the same time, the *clients* of prostitutes are rarely mentioned as risks.

[135]

- While IV drug users are defined as a "group" because they engage in a specific set of illegal behaviors, Stephanie Kane and Theresa Mason observe that in contrast the risk-group category "sexual partner" lacks correspondence with "any shared social scene and identity."[10] "Sex partners" are not only unlikely to constitute a real-life group, individually its members may be unaware that their partners and thus they themselves are at risk for HIV. And as for self-perceived identity, who thinks of themselves as a "sex partner" anyhow? ("That was no sex partner, that was my wife.")

- The sexual partners of people with hemophilia and blood disorders, for the most part women, are primarily dependent for information on their male partners (the patients); but as Patrick Mason and colleagues report, these men are often determined not to let the disease interfere with their lives, thus sometimes exclude their partners from opportunities to gain medical information about hemophilia, AIDS, and HIV risks.[11]

- To classify "mothers of pediatric AIDS patients" as a risk group is problematic for several reasons. First, the identity Mother in AIDS discourse regularly erases that of Woman. In one review article on psychological issues in AIDS, women were mentioned only under such headings as "pediatric AIDS"; a 1992 *New York Times* article described efforts to prevent transmission of the virus from mother to fetus, expressing little interest in the mother's fate. Second, the category suggests that the woman is "at risk" because she is the mother of a baby with AIDS, as though she acquired HIV from the baby. Finally, most seriously, her identity as mother enables her HIV status to be identified through legal hospital testing of newborn infants, thus depriving her of the right to refuse HIV testing still granted to other citizens.

Over the long run, confusions have clouded the terminology of both popular and research literature:

- Some researchers define any single, unmarried, sexually active woman as a prostitute or establish no clear distinction between "sex with multiple partners," "promiscuity," or "prostitution"; some lump together "sexual contact with multiple partners, including prostitutes." Contrast this with the identity of Mother just discussed: as Cindy Patton has observed, a woman in AIDS discourse has either a uterus or a vagina but never both at the same time.[12]

• The term "heterosexual AIDS" is especially confusing because it's used to refer both to cases of AIDS attributed to heterosexual transmission of HIV and to cases of AIDS among people identified as heterosexual (e.g., many infected by transfusions, needle sharing, etc.). Moreover, "heterosexual transmission" is sometimes defined so that it exempts heterosexual people of color, heterosexuals who are poor, sick, shoot drugs, or live in Africa (Haiti, Miami, New Jersey, etc.). Infected non-U.S. women are assumed to have engaged in something other than "ordinary sexual intercourse"; infected U.S. women of color are assumed to be drug users; infected U.S. white women are assumed to be flukes. The bottom line: educated white middle-class heterosexual people with Cuisinarts do not get AIDS.

• Popular and media images of women and AIDS are marked by the same stock characters and stereotypes; exceptions are given unwarranted visibility; real women at risk, real women with HIV and AIDS, are invisible or alibied. These contradictions are particularly acute in media images directed explicitly toward "women at risk" and "women with AIDS." Young women of color in U.S. urban settings are one of the groups most in need of AIDS education; yet they have rarely been written about or pictured except in ways that reproduce stereotypes of passivity, ignorance, and irresponsibility. While images of white middle-class women send the important message that, despite stereotypes about the epidemic, many kinds of people are potentially at risk, they also reinforce the incorrect message to women of color that they are *not* at risk.

• On the one hand, women are termed "inefficient" and "incompetent" transmitters of HIV, the weakest link in the chain of transmission; on the other hand, especially in reference to female IV drug users, women of color, and women from countries where heterosexual transmission is common, women are transformed into "reservoirs," "vessels," "vectors," and "carriers" of infection who can presumably now transmit the virus very efficiently indeed.

• Apparent heterosexual transmission in Africa is repeatedly attributed to a variety of cultural practices of the "Other": unadmitted homosexual or "quasi-homosexual" transmission, unadmitted drug use, the practice of anal intercourse as a method of birth control, the widespread use of unsterilized needles, a history of immune suppression and infectious disease, scarification, clitoridectomy, prostitution, promiscuity, circumcision (its presence in females and absence in males), and violent, excessive, or exotic sexual practices. To account for the existence of widespread HIV infection among heterosexual men and women in Central Africa, for example, Gould in his *Cosmo* article offered two explanations: first, he asserted, homosexuality among African men is common but taboo and therefore not acknowledged to investigators; second, "many men in Africa take their women in a brutal way, so that some heterosexual activity regarded as normal by them would be closer to rape by our standards."[13] Like other stereotypes that have guided the interpretation of AIDS data, these supposed cultural differences have both a practical and a symbolic role in establishing and maintaining divisions between "us" and "them."

To sum up, longstanding identity of AIDS as a "gay disease" and a "man's disease" has the effect of placing the burden on women themselves to prove their own significance—as spokespersons, as persons at risk, as objects worthy of scientific and medical inquiry, and as agents of social justice and political change. For all the reasons I have noted, this has proved very difficult to do. The negative consequences are material and immediate: women encounter barriers to diagnosis and care, exclusion from treatment and social support programs, lack of information about sexuality and reproduction, lack of preventive technologies designed for women, and lack of resources and support services for women, children, and families. There are long-term consequences as well. Even as the AIDS crisis reveals the unreliability of everyday categories, those categories are being further codified in policies and regulations for special classes of women, including sex workers, poor women, lesbians, women in Third World countries, IV drug users, inmates, and childbearing women at large. The AIDS epidemic fuels a conservative agenda for women—marriage, family, children—and amplifies already vocal calls for protection and surveillance. For women, this includes court-ordered caesarean sections, penalties for health-care providers who provide contraceptive information, incarceration for drug-using pregnant women, and an end to legal abortion. Electing a new president and a different political party will not erase many years of successful conservative efforts—that will require organized collective political responses by and for women.

Fortunately, such a response appears to be underway at last. A number of projects suggest strategies for creating a strong political voice for women and, in the process, identities that enable them to address the AIDS epidemic effectively. One of the clearest examples was provoked by Gould's 1988 *Cosmopolitan* article. Women in the New York chapter of the activist organization AIDS Coalition to Unleash Power (ACT UP) identified errors of fact, flawed assumptions, outdated statistics, and claims contradicted by their own knowledge and experience. To challenge Gould's misleading advice to women they began with their knowledge that women *were* infected with HIV and *were* dying of AIDS. Picketing the offices of *Cosmopolitan*'s New York publisher, they asserted that "The *COSMO* girl CAN get AIDS," distributed flyers countering Gould's claims with extensive documentation, and urged the public to just "Say No to Cosmo" by boycotting the magazine. To follow up, the women in ACT UP produced and distributed a documentary video called *Doctors, Liars, and Women: AIDS Activists Say No to* Cosmo, which recorded the debate as it continued in local and national media; they formed the ACT UP Women's Caucus, which organized subsequent actions; and they produced the book *Women, AIDS, and Activism.*[14]

The *Cosmopolitan* action marks a significant step forward in identifying the AIDS epidemic as an urgent women's health problem, as a significant social and political crisis, and as a premier symbolic battleground of our times where battles today shape policies that will affect us all tomorrow. *Doctors, Liars, and Women* proclaims women's right to represent themselves and tell their side of the story: *Women, AIDS, and Activism* places the AIDS epidemic within a broad feminist framework, relates AIDS issues to ongoing scientific, clinical, and social issues for women, and spells out the epidemic's negative impact on civil rights, equality in the workplace, childbearing and reproductive freedom, and other areas.

Other projects suggest a range of approaches and possibilities:

- A growing body of productions from many sources—feminist and lesbian films and videos, photographs, posters, books and booklets, telenovelas, publications, comics and cartoons—offer many potential positions of identification and inscribe women in AIDS discourse in multiple ways.

- Detailed reviews of laws and policies related to women and AIDS illuminate an astonishing diversity of emerging problems as well as the consequences produced by the patchwork of local and state "solutions"—including newly invigorated criminal penalty laws for female sex workers and intensified legal vulnerabilities of many other women in many states.

- Taking a cue from Cindy Patton and Janis Kelly's groundbreaking *Making It: A Woman's Guide to Sex in the Age of AIDS*, some projects attempt to rethink eroticism. As Lynne Segal has written, if sex is constructed, then it can be reconstructed: "We need to use our imaginations to explore the edges of safety, to redefine danger, to engage fantasy, to develop play in the arena of sex now marked by this epidemic."[15] Explicitly addressing problems of communication surrounding sex, prevention, and risk-reduction practices as well as "how to" questions of everyday life, one project uses informal at-home gatherings modeled along the lines of Tupperware parties to encourage women to talk frankly about problems and solutions. Other projects foster communication among interested women through support groups, church gatherings, journals, newsletters, radio and television talk shows, or electronic networks.

[140]

- Some projects target specific communities. A workshop organized in 1989 addressed issues affecting "women in the hemophilia community," beginning the transformation of an official "risk group" (sex partners of men with hemophilia) into a lived identity and real community; as Suzanne Broullon reports, the Women's Outreach Network of the National Hemophilia Foundation (WONN) is now a national organization with local chapters across the country.

- A growing number of projects strive to cross divisions of race, sex, class, and ethnicity. For example, the female rap group TLC address their music videos, interviews, and explicit messages about self-esteem and safe sex to an audience of African-American women; the African-American community is also the target of two women in Houston, known as "the AIDS ladies," who contact drug users on the street at the same time that they work through women's church groups to reach a wider audience; the video *DiAna's Hair Ego* by Ellen Spiro documents DiAna's transformation of her beauty salon into a place where information and conversation about HIV, AIDS, and safer sex are readily available. Efforts are also attempting to link U.S. AIDS activists with women's groups in other countries, particularly in the Third World where women are subject to widespread surveillance and repressive legislation with fewer protections available.

- High-profile, media-savvy women's health organizations like the Women's Action Coalition (WAC) and the Women's Health Action Mobilization (WHAM) link the AIDS epidemic to specific struggles in women's health. Arguing that what happens with AIDS will influence future policy and practices in many arenas, they are adapting the strategies of ACT UP to fight for women's access to experimental drugs and the need for more and better research and treatment of breast cancer. Similarly, women are lobbying for increased reproductive and molecular research on diseases that disproportionately affect women.

[141]

At the heart of the joke about Joe and Jane with which I opened this essay is the question of identity, a complicated phenomenon we know less about than we think we do. Problems of identity—problems, that is, of whether or not something is "the same as" something else—underlie many seemingly unrelated issues with regard to women and AIDS/HIV. As I have tried to suggest here, the fluidity, ambiguity, and questioning posited by the joke—not to mention the brutal social reality embodied by its punch line—are absent from most conceptions of gender that inform AIDS discourse, whether in medical journals, epidemiological surveillance reports, or popular culture. As I said at the outset, an effective response to an epidemic (as to any widespread cultural crisis) depends on the existence of identities for whom that epidemic (or crisis) is meaningful—and stories in which those identities are taken up and animated. I have also sought to show how identities and narratives about women and AIDS have tended to discourage an effective response. Although many identities for women are readily available in relation to AIDS, they are rarely useful or meaningful. Despite sporadic efforts at clarification, the question of gender and AIDS/HIV remains a problematic component in most domains of AIDS discourse, where established conceptions of women shape a cycle of research, representation, and analysis that perpetuates a view of AIDS as a man's disease and discourages sustained focus on women's issues. In the culture at large, flourishing stereotypes generate simultaneous visions of women as impervious to infection and diseased fantasy figures.

With a few notable exceptions, I have argued, women and AIDS is also a problematic component of feminist discourse. While many women's magazines and feminist journals have dutifully published their "what women should know about AIDS" articles, the epidemic is still represented rather narrowly as a personal health risk which basic precautions can virtually eliminate. Faced with inconclusive data, conflicting media reports, and increasingly difficult efforts to preserve a working definition of sisterhood and feminist identity, women remain confused and, despite the rising body count, reluctant to add AIDS to an already overburdened agenda. Put another way, in feminist discourse the categories "women with AIDS" and "women with HIV" languish in the divide between foundationalist and constructionist accounts of gender.

What can we do? One challenge before us is to acknowledge the legitimacy and usefulness of treating identity as complicated, as something to be investigated rather than assumed. At the same time, we must take responsibility for the growing social, material reality of women with HIV and AIDS, a category strikingly unified by its predictable and negative social consequences. In the process of becoming Jane, Joe's identity may be negotiable; but when Jane picks up her paycheck, negotiation stops. These social consequences constitute another challenge we must reckon with. Finally, we must assert that the AIDS epidemic is a premiere symbolic battleground where war will be waged incessantly, where language and reality will continue to shape each other, where we can see the health-care system in action and work to change it, and where women's futures will in part be determined. Informed by a feminist analysis that takes its historical and cultural context seriously and specifies its tasks carefully, we must come to see the story of women and AIDS as a dense narrative about women's health and American society; economic opportunity and political power; sexuality and safety; law and transgression; individual autonomy and reproductive freedom; the right to social services and health-care resources; the deformities of the American health-care system; alliances with others; and about the significance of identity—including our own—in everyday life.

Notes

Elaine Scarry

1. For the elaboration of this argument, see Elaine Scarry, "Consent and the Body: Injury, Departure, and Desire," *New Literary History* 21, no. 4 (Autumn 1990), pp. 867–896.
2. Gilbert Ryle, *The Concept of Mind* (London: Hutchinson, 1949), p. 63.
3. William James, *The Principles of Psychology*, vol. 2 (New York: Dover, 1950), p. 486.
4. Augustine, *Concerning the City of God against the Pagans*, trans. Henry Bettenson (Harmondsworth: Penguin Books, 1984), Book 14. 24, pp. 587–588.
5. Epictetus, "Discourses," in *Epictetus: The Discourses as Reported by Arrian, the Manual, and Fragments*, trans. W. A. Oldfather, Loeb Classical Edition, vol. 2 (Cambridge, Mass.: Harvard University Press, 1985), pp. 4.1.34.
6. Ryle calls attention to the way the tradition perceives in volitions what I am calling here their "artifactual" nature. Ryle himself, in contrast, is arguing against the usefulness of the construct of "the will"; he designates volitions "artificial" in the negative sense of contrived or false.
7. James, *The Principles of Psychology*, vol. 2 pp. 568–569.
8. Augustine, *The City of God*, Book 14. 24, p. 588.
9. Richard Selzer, *Mortal Lessons: Notes on the Art of Surgery* (New York: Simon & Schuster, 1974), pp. 93–94.
10. Ibid.
11. Ibid.
12. This perceptual habit in Donne, Leonardo, Nicolas of Cusa, as well as Renaissance physicians, is examined in my essay "Donne: But yet the body is his booke," in *Literature and the Body: Essays on Populations and Persons*, ed. Elaine Scarry (Baltimore: Johns Hopkins University Press, 1988), pp. 70–105.
13. Heinrich von Kleist, "On the Marionette Theater," in *German Romantic Criticism*, ed. A. Leslie Willson (New York: Continuum, 1982), p. 240.
14. Charles Dickens, *Our Mutual Friend*, ed. Stephen Gill (New York: Penguin Classics, 1971), pp. 90–91.
15. Ibid., pp. 650, 717.
16. Ibid., pp. 554–555, 559.

17. Ibid., pp. 289, 294, 492, 497, 800.
18. Ibid., p. 495.
19. Charles Baudelaire, "A Philosophy of Toys," in *The Painter of Modern Life and Other Essays*, trans. Jonathan Mayne (New York: DaCapo Press, 1964), p. 198.
20. Homer, *Iliad*, trans. A. T. Murray, Loeb Classical Edition (Cambridge, Mass.: Harvard University Press, 1924), p. 18, line 420. See also Robert Fagles's beautiful translations of the various passages cited here: Thetis is, for example, described as "glistening footed."
21. *New Larousse Encyclopedia of Mythology*, intro. Robert Graves (New York: Putnam, 1968, c. 1959).
22. Plato, *Euthyphro*, trans. Hugh Tredennick, in *Last Days of Socrates* (New York: Penguin, 1969), pp. 34, 40.
23. *The Book of Counsel: The Popol Vuh of the Quiche Maya of Guatemala*, ed. Munro S. Edmonson, Middle American Research Institute, Publication 35 (New Orleans: Tulane University, 1971), Book 1, pp. 183–186. See also Book 11, pp. 184, 350, 506, 712.
24. All the engineering accounts are hymns of praise to the miraculous motion of the biological arm, which, like a wing, is almost never still, and which presents almost insurmountable difficulties for mimesis.
25. Douglas Lamb and Hamish Law, *Upper-Limb Deficiencies in Children: Prosthetic, Orthotic, and Surgical Management* (Boston: Little, Brown, 1987), p. 64.
26. Marian Weiss, et al., *Myoplastic Amputation, Immediate Prosthesis, and Early Amputation*, Warsaw and Konstantin Rehabilitation Clinic (Washington, D.C.: GPO, 1971), p. 76.
27. Lamb and Law, *Upper-Limb Deficiencies in Children*, p. 83.
28. One occupational group very prone to accidents involving the limbs is farmers. The Purdue School of Engineering publishes a newsletter, *Breaking New Ground*, that is sent to farmers: it contains accounts of new prosthetic inventions and techniques and thus serves as a crucial source of information for those who need to continue farming after they have lost a natural limb.
29. Colin Murray Parkes, "Psycho-Social Transitions: Comparison between Reactions to Loss of Limb and Loss of a Spouse,"

British Journal of Psychiatry 127 (September 1975), pp. 204–210.

30. Rainer Maria Rilke, "Reflection on Dolls," in *Where Silence Reigns: Selected Prose,* trans. G. Craig Houston (New York: New Directions, 1988), p. 44.

31. Ibid., p. 46.

32. Ibid., p. 43.

33. Ibid., p. 45.

R. C. Lewontin

1. See, for example, Committee on Mapping and Sequencing the Human Genome, *Mapping and Sequencing the Human Genome* (Washington, D.C.: National Academy Press, 1988); Daniel J. Kevles and Leroy Hood, eds., *The Code of Codes: Scientific and Social Issues in the Human Genome Project* (Cambridge, Mass.: Harvard University Press, 1992); Jerry E. Bishop and Michael Waldholz, *Genome: The Story of the Most Astonishing Scientific Adventure of Our Time—The Attempts to Map All the Genes in the Human Body* (New York: Simon & Schuster, 1990); Lois Wingerson, *Mapping Our Genes: The Genome Project and the Future of Medicine* (New York: Dutton, 1990); Joel Davis, *Mapping the Code: The Human Genome Project and the Choices of Modern Science* (New York: Wiley, 1991); Christopher Wills, *Exons, Introns, and Talking Genes: The Science Behind the Human Genome Project* (New York: Basic Books, 1991); Dorothy Nelkin and Laurence Tancredi, *Dangerous Diagnostics: The Social Power of Biological Information* (New York: Basic Books, 1989); David Suzuki and Peter Knudtson, *Genethics: The Ethics of Engineering Life* (Cambridge, Mass.: Harvard University Press, 1990); and Daniel J. Kevles, *In the Name of Eugenics: Genetics and the Uses of Human Heredity* (New York: Knopf, 1985).

2. James Dewey Watson, "A Personal View of the Project," in *The Code of Codes,* ed. Kevles and Hood, p. 165.

3. Walter Gilbert, "A Vision of the Grail," and Watson, "A Personal View," in ibid., pp. 83–97, 164–173.

4. Suzuki and Knudtson, p. 163.

5. Remarks made at the First Human Genome Conference in the Human Genome Project in October 1989. Quoted by Keller in "Nature, Nurture, and the Human Genome Project," in *The Code of Codes,* ed. Kevles and Hood, p. 282.

6. Ruth Schwartz Cowan, "Genetic Technology and Reproductive Choice: An Ethics for Autonomy," in *The Code of Codes,* ed. Kevles and Hood, pp. 244–263.

7. Pressure against the paper was also brought by scientists in the genome sequencing establishment on the editor of the journal in which it was to be published, including one of the contributors to *The Code of Codes.* As a result, the editor delayed its publication, demanded changes in galley proofs, and asked two defenders of the method to write a counterattack. One report of the scandal is given in Lesley Roberts's "Fight Erupts over DNA Fingerprinting," *Science* 20 December 1991, pp. 1721–1723.

8. Based on *Frye* v. *United States* 293 F: 2nd D.C. Circuit 1013, 104 (1923).

9. Committee on DNA Technology in Forensic Science, *DNA Technology in Forensic Science* (Washington, D.C.: National Academy Press, 1989). The reader should know I am not a disinterested party either with respect to the report or to the body that sponsored it. I have twice testified in federal court on the weaknesses of DNA profiles, am the author of a position paper that was a basis for the original very critical version of the NRC report's chapter on population considerations, and am the author, with Daniel Hartl, of a highly critical paper in *Science* that was the object of considerable controversy. I resigned from the National Academy of Sciences in 1971 in protest against the secret military research carried out by its operating arm, the National Research Council.

10. See M. Allison, "The Radioactive Elixir," *Harvard* (January–February 1992), pp. 734–775.

Paula A. Treichler

1. This text is condensed and adapted from my essay "Beyond Cosmo: AIDS, Identity, and Inscriptions of Gender"; the longer version, with illustrations and full citations, appears in *Camera Obscura* 28 (1992), pp. 21–76. An abbreviated version also appears in

[145]

Margot Lovejoy, *The Book of Plagues* (Philadelphia: Lori Spencer, Borowsky Center for Publication Arts, 1994).

2. Gina Kolata, "Ideas and Trends: Who is Female? Science Can't Say," *New York Times* 16 February 1992.

3. P. Clay Stephens, "U.S. Women and HIV Infection," in *The AIDS Epidemic: Private Rights and the Public Interest*, ed. Padraig O'Malley (Boston: Beacon, 1988), pp. 381–401.

4. Robert E. Gould, "Reassuring News About AIDS: A Doctor Tells Why You May Not Be at Risk," *Cosmopolitan* 204, no. 1 (January 1988), p. 146.

5. John Langone, "AIDS: The Latest Scientific Facts," *Discover* (December 1985), pp. 28–53.

6. John Langone, *AIDS: The Facts* (Boston: Little, Brown, 1988).

7. See R. R. Redfield, et al., "Heterosexuality Acquired HTLV-III/LAV Disease (AIDS-Related Complex and AIDS): Epidemiologic Evidence for Female-to-Male Transmission," *Journal of the American Medical Association* 254 (1985), pp. 2094–2096; R. R. Redfield, et al., "Female-to-Male Transmission of HTLV-III," *Journal of the American Medical Association* 255 (1986), pp. 1705–1706; and Harold Sanford Kant, "The Transmission of HTLV-III" (letter to the editor), *Journal of the American Medical Association* 254 (October 1985), p. 1901.

8. Data reported here is from the 1993 year-end edition of the Centers for Disease Control's quarterly *HIV/AIDS Surveillance Report*. Because of extensive changes in the CDC's AIDS/HIV reporting system and software, actual publication of official year-end data for 1993 and for the first two quarters of 1994 are expected in fall 1994; numbers reported here, based on AIDS cases reported through December 31, 1993, are termed provisional by the CDC. Some of the categories I mention here have been modified; in addition, the CDC now reports data on HIV infection from states with mandated HIV reporting. Readers may obtain the CDC's surveillance reports by calling the CDC National AIDS Clearinghouse at (800) 458-5231.

9. *Newsweek* 18 November 1991, p. 59.

10. Stephanie Kane and Theresa Mason, "'IV Drug Users' and 'Sex Partners': The Limits of Epidemiological Categories and the Ethnography of Risk," in *The Time of AIDS: Social Analysis, Theory, and Method* (Newbury Park, Calif.: Sage, 1992), pp. 199–222.

11. Patrick J. Mason, Roberta A. Olson, and Kathy L. Parish, "AIDS, Hemophilia, and Prevention Efforts Within a Comprehensive Care Program," *American Psychologist* 43 (1988), pp. 971–976.

12. Cindy Patton and Janis Kelly, *Making It: A Woman's Guide to Sex in the Age of AIDS* (Boston: Firebrand, 1987; 2d ed., 1992).

13. Gould, "Reassuring News about AIDS," p. 146.

14. ACT UP/NY Women's Book Group, *Women, AIDS, and Activism* (Boston: South End Press, 1992).

15. Lynne Segal, "Lessons from the Past: Feminism, Sexual Politics, and the Challenge of AIDS," in *Taking Liberties: AIDS and Cultural Politics*, ed. Erica Carter and Simon Watney (London: Serpent's Tail, 1989), pp. 133–145.

[146]

III

Information,

Artificiality,

and Science

Making Sense Out of Nonsense: Rescuing Reality from Virtual Reality

Gary Chapman

I thought I'd start off by telling you some stories—true stories—about computers.

In June 1985 the space shuttle performed one of its first missions for the Star Wars program. The shuttle was to fly over a mountain in Hawaii, on top of which was mounted a high-powered laser. The laser was supposed to fire up through the atmosphere when the shuttle flew over, and the beam was going to be reflected back to Earth, for measurement, by a huge mirror mounted to the shuttle's underside. Now, many of you know that the shuttle isn't "flown" by a pilot—it's driven by computers, completely preprogrammed: if there's any change in the mission, the new instruction codes are transmitted from the ground. The programmers of this software had standardized every measure of distance in nautical miles—but they forgot to recalculate the height of the mountain. So, instead of looking for a mountain that was ten thousand *feet* high, the shuttle's computers searched for a mountain that was ten thousand *nautical miles* high, which is much higher than the shuttle's orbit. So, when the shuttle flew over the appointed spot, it flipped over, pointing the mirror out into space—and the laser hit the top of the shuttle. Nothing happened; the 15 million-dollar experiment was a failure.

The next story is about a young boy in the late 1960s in Palo Alto, California, who often went to a Baskin Robbins ice cream store. At the time, Baskin Robbins ran a promotional scheme in which kids filled out a little card, and when their birthday rolled around they'd get a coupon in the mail for a free ice cream cone. Well, this boy got a big stack of these cards and filled them out with the names of fictional boys with birthdays throughout the year, all living at his address. He got free ice

[149]

cream cones throughout the year. But ten years later all these nonexistent boys started getting mail from the Selective Service Administration saying that they had failed to register for the draft and were subject to criminal prosecution. Baskin Robbins was embarrassed into admitting that it had been selling this database to the Pentagon for years, and it discontinued the practice.

Now, one last, quick story: the head, or "primate," of a certain Catholic denomination in Canada recently received in the mail an invitation to join his fellow primates at the National Canadian Primate Research facility!

Now, what do these stories have in common, other than being funny and involving computers? All involved transactions that resulted in a kind of comical nonsense—and all had some material effects in the world.

The standard explanation for such things is that computers don't have any common sense: they don't "just know" that there are no ten thousand–mile mountains or that churches aren't headed by monkeys. But this mischaracterizes what computers are about because it implies that they might be given common sense some day, or that they might now have some kind of sense other than common sense, so, we merely need to compensate for this by admiring the "sense" they *do* have. This is misleading. I'd like to present another model for the computer, one that we need to assert.

When we talk about a computer we're talking about a box filled with various kinds of electronic switches and motors, which perform particular tasks according to encoded sets of instructions. This isn't an entirely accurate view of what a computer is or does, though, because it ignores a very important aspect of the computer: it is also a social arti-

[150]

fact that mediates a series of social interactions conceptualized as a model *through* these encoded instruction sets. To clarify: human beings conceptualize this model, others encode it in a formal language, still others actually use computers, and still others, even further down the line, are subject to what the computer does. Let me give you a concrete example, a real tragedy that illustrates this well. In 1989 the United States was sending warships to the Persian Gulf to protect oil shipments.

One frigate, the USS *Stark*, became famous when it was hit by two Iraqi-fired but French-made missiles; thirty-seven sailors died. The *Stark* had a sophisticated, computerized air defense system aboard, which is monitored by operators who watch a round screen inscribed with concentric circles. You would likely think that it would show a spatial representation of the missile's *proximity*, but in fact the missiles shown are ranked by their order of hostility: the more hostile they are, the closer to the center of the screen they appear. The rationale behind this counterintuitive arrangement is that, in a very intense battle with lots of missiles flying around, the fire-control officer needs to prioritize defenses against not missiles per se but against *hostile* missiles. But these Iraqi-fired missiles never showed up on this screen—they were identified visually, which gave the sailors about six seconds to react.

About two weeks after the *Stark* was hit, the navy changed the software in all air defense systems similar to the one on the *Stark*. Shortly afterward, the commander of the *Stark*—who was run out of the navy—speculated that his air defense system didn't work because the French-made missiles were classified as "friendly"—built by an ally, part of the NATO stockpile. Moreover, there's some indication that the Iraqis had modified the missiles in order to penetrate Iranian air defenses, because the Iranians were using U.S.-made missiles, which our government had covertly sold to them.

It is interesting how these various things are bound up with each other, but the point is this: these sailors' lives hinged on the software parameters programmed by someone who thought he knew what this ship was going to run into long before these events. So, a whole chain of human events was mediated by what boils down to a box of electronics.

Thus, when we talk about what goes on in a computer, we're talking about an entire complex of relations, assumptions, actions, intentions, design, error, too, as well as the results, and so on. A computer is a device that allows us to put cognitive models into operational form. But cognitive models are fictions, artificial constructs that correspond more or less to what happens in the world. When they don't, though, the results range from the absurd to the tragic.

It's a familiar thesis, that tools and machines embody conceptual models, and so embody certain values we hold about the world. A dense example: the difference between a short-handled hoe and a long-handled hoe, aside from the obvious difference in the length of the handle or the physics of its use, is the fact that a short-handled hoe forces one to bend over. This has political implications, which are being fought over as we speak in the American South and in Mexico: laborers are being forced to use short-handled hoes so that foremen can see from a distance who's working and who isn't. Two conceptual models—one emphasizing ergonomic efficiency, the other political efficiency—are in conflict over the length of a hoe handle, over the conceptual models according to which our tools are configured.

Computers are unique in that their function is *entirely* determined by the model programmed into them: they're the "universal machine." Not only can they simulate any other machine but they are completely reconfigurable, depending on the instruction sets programmed into them. On the other hand, they can produce what can only be called nonsense—*while working exactly as they're supposed to*. More than non-sensical, though, these results can be catastrophic. Computers, as we all know, are enmeshed in the control of our nuclear arsenal, and there is some danger that an error in one of these computer systems could end life on the planet. The computer will have performed exactly as it was programmed to.

I stress this kind of extreme because, in this light, the category "nonsense" becomes a bit more alarming. A model that "works" in a computer—it is operational and plausible, in other words, algorithmi-cally correct—can certainly be wrong in many other ways: it might be offensive; it might, over time, affect our behavior and assumptions in ways detrimental to our fundamental human interest. But because the computer is often seen as a kind of value-free device, we tend to view its commands as being other than human-derived. We tolerate from computers the imposition of rules and power that we would never tol-erate if it were proposed directly by another human being. When we hear about a computer that's made a monkey out of a priest, we laugh about it, but we're typically silent when computers degrade us and our abilities, invade our privacy, and threaten our security by controlling weapons of mass destruction. We're taught to view the computer as a machine that performs various tasks involving symbol and numerical analysis, and we're taught that computers sometimes "make mistakes," but we aren't taught that the models according to which they're programmed reinforce corporate, military, and class power—as do all other human-made artifacts. These models are typically viewed as exempt from examination and challenge. Yet computers, particularly—I think because of their curiously all-purpose instrumental character—deform the human character to such an extent that the very ability to recognize the artificial nonsense of these systems and models is extinguished: all that's left to affirm our own human uniqueness and our critical capability is the smug assertion that a computer is stupid enough to look for a ten thousand–mile high mountain.

Reforming the way that we look at computers, then, will involve reforming cognitive models throughout society. This isn't to imply that computer technology is value-neutral, or that only the use to which it is put indicates its normative character. Computers *do* embody values— in their cost, their material makeup, their design, the fact that they exist in our economy as a commodity, and so on. And, of course, they tend to support an overvaluation of quantifiable data, a tendency that has many social effects. But none of these characteristics is fixed permanently in the computer as such—so the particular cognitive operational models that we program into computers are subject to some political contestation. Some people think that computer technology is not susceptible to reform. This is antipolitical and reactionary: computers can support progressive and humane values.

I work with an organization, Computer Professionals for Social Responsibility, which has established a new program, the 21st Century Project. Our goal is a comprehensive, national campaign to redirect government funding for science and technology away from military research and development toward peaceful and environmentally responsible ends. But it's not enough to simply redirect money from the military to civilian programs, because civilian programs will embody the same values that military programs do, merely with a different mission. We have to rethink civilian programs. The 21st Century Project is working toward uses of computer technology that don't replace workers, reinforce government power, provide public and private surveillance systems, or anesthetize people into robotic consumers. Computers can serve to support skills and workers, give voice to people in their communities and in government, build communication systems that enhance democratic participation, and stimulate people in creative ways—but first we have to demand that these values be reflected in government funding priorities that will support new research and development in information systems and other technology.

One project we're working on is an attempt to integrate the reform of national policy grassroots–level activism around a Defense Department research facility in Austin, Texas, called Semitech, which orchestrates research in semiconductors. Semitech, which is a fairly unique government-industry collaboration—supported by the government to the tune of $100 million per year and by the fourteen largest U.S. semiconductor companies—is located in a very poor, largely Hispanic and African-American community in East Austin. When we began to organize this community, we discovered that the people who live around this federal facility knew absolutely nothing about it. Nor did they know that the semiconductor industry has a terrible record of polluting groundwater resources, or even what semiconductors do, or why this facility is crucial to the national strategy of economic and military domination. So we organized the commu-nity and demanded that the community now be involved in Semitech's planning and development. We've asked Congress, in the legislation that reauthorizes Semitech, to incorporate community and environmental representation on Semitech's board, a ten-percent earmarking of Semitech's federal funds to researching environmentally sound practices in semiconductor manufacture, a right-to-know clause for the people who live around the facility, and so on.

The 21st Century Project will also convene four national working groups to devise the beginnings of a new agenda for science and technology policy. These four working groups will, over a period of about a year, be looking at opportunities for new investment in science and technology around four particular issues. First, democratizing the science and technology development process away from corporate and academic research labs and toward institutions that will permit citizens a say, and more of a stake, in the development of the country's material infrastructure. Second, promoting "sustainable development," that is, methods of industrial production that won't foul the environment, by reconceptualizing the production process and the materials it uses. Third, research and development into the nature of computer design and use, in order to enhance and develop people's skills instead of replacing them with "participatory design" and "skills-based automation." And, finally, a reevaluation of the development of computer and communication infrastructure, which the United States is right now beginning to build—the backbone of the new communications system that will replace Internet and revolutionize communications in the United States. We're trying to make sure that the new infrastructural networks permit community participation, public access, and democratic means of deciding policy, rather than perpetuating domination by military and corporate interests. Following the working groups' development of these plans, we will launch a public education campaign to educate people about alternative ways of deploying technology in our society.

The 21st Century Project has three long-term goals we want to build into the nation's science and technology policy: ending the continual refinement of weapons and, ideally, shutting down the arms trade altogether; creating systems of sustainable and equitable production to preserve the environment; and giving priority to the full development of human potential through work by reexamining the ways in which we use technology to support human skills.

I hope that you'll hear more about it in the future; we're trying to engage citizens at the grassroots level and change national policy at the same time.

What Do Cyborgs Eat?
Oral Logic in an Information Society

Margaret Morse

> "Well, I'll eat it," said Alice, "and if it makes me grow larger,
> I can reach the key; and if it makes me grow smaller, I can
> creep under the door; so either way I'll get into the garden
> and I don't care which happens!"
>
> —Lewis Carroll, *Alice in Wonderland*[1]

For couch potatoes, video game addicts, and surrogate travelers of
cyberspace alike, an organic body just gets in the way.[2] The culinary
discourses of a culture undergoing transformation into an information
society will have to confront not only the problems of a much depleted
earth but also a growing desire to disengage from the human condition.
Travelers on the virtual highways of an information society have, infact,
at least one body too many—the one now largely sedentary carbon-
based body at the control console that suffers hunger, corpulency, illness,
old age, and ultimately death. The other body, a silicon-based surrogate
jacked into immaterial realms of data, has superpowers, albeit virtually,
and is immortal—or, rather, the chosen body, an electronic avatar
"decoupled" from the physical body, is a program capable of enduring
endless deaths. How can organically embodied beings, given these physical handicaps, enter an electronic future? Like Alice, this requires asking ourselves if and what to eat.

Some theorists in future-oriented subcultures who have wholeheart-
edly embraced technology (or who, as critics, at least speak from its
belly) have posed the union of machine and organism as the hybrid
meld, the *cyborg*, a "human individual who has some of its vital bodily
processes controlled by cybernetically operated devices."[3] However satis-

[157]

fying such an imaginary blend might be, the actual status of the cyborg is murky as to whether it is a metaphor, a dreamlike fantasy, and/or a literal being; and its mode of fabrication and maintenance is, practically at least, problematic.

Consider such a mundane and practical problem as this: What do cyborgs eat? After all, the different nutritional requirements of silicon- versus carbon-based intelligence of the mammalian persuasion are not negotiable in material reality. The alimentary process and its beginning and end products, food and waste, tie us inextricably to the organic world. Both the need to eat and the pleasure of eating are part and parcel of the condition of mortality which electronics are spared. It is unlikely that the very notion of "breakfast" (or lubrication cycle? power feed?) would have much meaning for the relatively immortal and virtual parts of a creature, which might suffer obsolescence, silica fatigue, and sudden crashes but not hunger or death. Willing the cyborg into being appears to be the equivalent of wishing the problems of organic life away. Yet unless the human is erased entirely, food and waste will enter the cyborg condition.

The more immediate question then is: What do humans who want to become electronic eat? For we are no longer talking about metaphors or electronic prostheses that extend organic body functions (in the way Marshall McLuhan understands the media, for instance), or even about Frankensteinian reassemblage or Tin Man–like displacements of the organic body part by part. In this more *mechanical* sense, cyborgs with heart monitors, organ implants, and artificial limbs already walk the earth. The contemporary fantasy is, rather, how, if the organic body cannot be abandoned, it might be fused with electronic culture in what amounts to an oral logic of *incorporation*.

[158]   In the first section of this admittedly speculative essay on food (or *nonfood*) in the context of body loathing and machine desire, I introduce the oral logic of incorporation into the electronic machine—for instance, that of eating/being eaten or of being covered by a second skin. The second section explores the contemporary socioeconomic context of famine and undesirable abundance as well as the ways in which "fast" and "fresh" food ideologies have failed their democratic promise. It is in this cultural context of ideological failure and the desire

to become (not merely to have) electronic machines that food per se can be—at least symbolically—refused. Indeed, body loathing entails food loathing, which manifests itself in food that negates its value as such (that is, as *nonfood*) as well as in other ways of purifying the organic body from "meat."[4] (When food itself is considered unhealthy, *nonfood* also has an odd relation to the discourses of health and nutrition.) The third and final section describes different modes of culinary and corporeal negation, including the psychological defense mechanisms of repudiation, denial, and disavowal. These defenses are means of purification from the organic associated with culinary phenomena in cyberpunk fiction, virtual reality, artificial fat, and smart drinks and drugs.[5]

In closing, I consider the inverse process, that of contaminating the electronic body and the virtual world with the organic—for example, by vomiting, as if turning the body inside out. Reversing the alimentary continuum is the mark of a largely misunderstood and recently controversial strand of "excretory" art. Smearing the body with food waste/simulated bodily fluids puts the inside on the outside, as if turning the body inside out in a symbolic rendering of *abjection*. Ultimately, the strategy of inversion also promotes a different cultural agenda. Insofar as it is the electronic body that is smeared, the electronic machine is enveloped in a second skin of human waste. That is, rather than making cyborgs by accommodating the organic to the electronic, the eater (electronic culture) becomes the eaten, in a symbolic initiation of the cyborg into the human condition.

I. Oral Logic: The Dialectics of Incorporation

> As she withered, sucked of energy, he became more alive and
> animated. When she brought our tea, her face was clouded and [159]
> dark, her shoulders bunched and turned in. He had eaten her
> alive. I sat amazed watching this psychic cannibalism.
>
> —William Patrick Patterson[6]

"Identification" as a mirrorlike relationship to visual media has been the dominant theoretical model for the construction of subjectivity for at least two decades. However, the model seems ill-suited to explain the "immersive" aspects of electronic media, which can be better understood in terms of oral logic. What was once thought as a stage of development in infancy to be relegated to the background and to fantasy in adult life, promises to become the dominant mode of subject construction in the age of information. While the process of identification associated with the cinema paradoxically depends on distance,[7] the fusion of oral incorporation is a more-than-closeness: it involves introjecting or surrounding the other (or being introjected or surrounded) and ultimately, the mixing of two "bodies" in a dialectic of inside and outside that also can involve a massive difference in scale. Bodies in oral logic can range from very small (usually, but not always, the eaten) to the immense (often, but not always, the eater). The body of the other can be as large as an intrauterine-stomachic-intestinal interiority or virtual void within which one is "immersed" (consider, for instance, electronic encapsulation of the body in virtual reality), or as small as a smart pill one ingests. There also appears to be a dialectic between eating/being eaten and the sucking out, piercing, and fragmenting of the body (as if into food) versus resurrecting it into wholeness or preserving it in an incorruptible state. Note that, unlike identification, incorporation does not depend on likeness or similarity or mirrors in order to mistake the other as the self; in an "oral-sadistic" or "cannibalistic" fantasy, the introjected object (electronic machine or human body as other, depending on who eats whom) is occluded and destroyed, only in order to be assimilated and to transform its host.

Eating

One method of cyborg construction is that of introjection and absorption. In the words of Laplanche and Pontalis, "the subject, more or less on the level of phantasy, has an object penetrate his body and keeps it 'inside his body': [Incorporation] means to obtain pleasure by making an object penetrate oneself; it means to destroy this object; and it means, by keeping it within oneself, to appropriate the object's qualities. It is this last aspect that makes incorporation into the matrix of introjection and identification."[8]

That is, *pace* Brillat-Savarin, the nineteenth-century gastronome, who you are not only decides what you eat, what you eat or introject also determines who you are. Therefore, in order to become a cyborg—a partly human, partly electronic entity—a human must eat the stuff of cyborgs. Indeed, cannibalistic fantasy plays a great part in this oral logic "marked by the meanings of *eating* and *being eaten*."[9] As J. G. Ballard's aphorism on "Food" proposes, "Our delight in food is rooted in our immense relish at the thought that, prospectively, we are eating ourselves."[10]

Currently, when we want to introject cyborgs, "smart" drinks and drugs will have to do. Built along the analogy of *smart* appliances, houses, and bombs, the adjective *smart* attributes some degree of agency and, at times, human subjectivity to the object world. "Smart" pill and powder cuisine consists of vitamins and/or drugs, laced at times with psychotropics and aimed directly for the brain.[11] To the cyberpunk culinary imaginary, these chemicals are decidedly utopian, a kind of lubricant or "tuneup"[12] for wetware that breaks the blood-brain barrier, makes neurons fire faster, and encourages dendrite growth, *not unlike* the networks linking the electronic channels along which information flows.

But the more fundamental, albeit speculative, answer for humans who want to transcend the organic body and its limits—that is, those who want to be cyborgs—is to eat *nonfood*, food that negates the very idea of the organic or "natural" value of food. Vitamin gels and chemical soups qualify precisely *because* they blur the categories of food and drugs, anticipating the advent of what futurologist Faith Popcorn calls "food-ceuticals."[13] Capsules of what are tantamount to brain chemicals condense "intelligence" into a magical essence or fetish for transforming the human brain into a high-performance electr(on)ic machine.

Smart drugs are chemically targeted at the brain—but they are considered efficacious at bringing the flesh at the console along for the ride. For instance, despite the fact that Durk Pearson of Durk Pearson and Sandy Shaw® reports spending "all my time lying flat on my back on my waterbed with my computer," he claims to have "good muscles," thanks to smart nutrition.[14] Thus the strategy is not only to feed the mind but in the process to *purify* the body of organic deterioration. For would the ideal cyborg, an electronic *kouros* or imaginary of machine/human perfection, have any need of flesh? To become cyborg, one does not eat the apple of the knowledge of good and evil, but something more like the body of the deity, the host of disembodied information.

Being Eaten

Although machines are spared the need for organic nourishment, the *fusion* of organic and electronic must also logically include the possibility of being eaten by electronic machines.[15] Some scientists of artificial intelligence anticipate such an event as ecstasy. For instance, Hans Moravec, author of *Mind Children*, foresees leaving the organic body behind like an empty shell after what amounts to having the brain scooped out, emptied, exhausted bit by bit by one's own advanced robot mind-child. Brain cells or wetware would be displaced with silicon, byte by byte:

> In a final, disoriented step the surgeon lifts its hand. Your suddenly abandoned body dies. For a moment you experience only quiet and dark. Then, once again, you can open your eyes. Your perspective has shifted. The computer simulation has been disconnected from the cable leading to the [robot brain] surgeon's hand and reconnected to a shiny new body of the style, color, and material of your choice. Your metamorphosis is complete.[16]

(This fantasy is elaborated at length in Harry Harrison and Marvin Minsky's 1992 science-fiction novel, *The Turing Option*.) "Downloading consciousness" into a computer—which, according to Moravec, will be available by the mid-twenty-first century—would simulate brain functions but at an incomparably faster speed. Gerald Jay Sussman, a professor at MIT, once reportedly expressed a similar desire for machine fusion as the wish for immortality:

[162]

> If you can make a machine that contains the contents of your mind, then the machine is you. The hell with the rest of your physical body, it's not very interesting. Now, the machine can last forever. Even if it doesn't last forever, you [notice this logical lapse or, perhaps, metalepsis] can always dump onto tape and make backups, then load it up on some other machine if

the first one breaks .... Everyone would like to be immortal ....
I'm afraid, unfortunately, that I am the last generation to die.[17]

Far more recently, Larry Yeager's confession of why he "fell for artificial life" expressed a similar desire to "live on inside the chips."[18] Yet,
in O. B. Hardison's concluding image of humanity's immersion in or
engulfment by the machine in *Disappearing through the Skylight*, remnants of the organic body are nonetheless retained within the greater
body of technology and silicon-based intelligence, much as the mito-
chondria within human cells remind us of our origin in the sea and
in asexual reproduction.

Compare these images of incorporation *within* machines—be it as
enthusiastic vision or warning—with the image of disembodied, artificial intelligence in the Romantic imagination; a miniature artificial
man, Goethe's homunculus, was a created product of mind kept in a
bell jar. His greatest desire was to dissolve himself in the ocean, perceived by the homunculus himself as a female realm of pure body. His
immersion expressed a kind of death wish of erotic fusion with undifferentiated nature itself. In contemporary discourse about the future,
with its various degrees of hostility to organic life, intelligence that
breaks its corporeal container is seen as simply joining its like in a great
digital sea of data. So, the virtual realm is tied symbolically to *immersion*
and all its attendant hopes for transcendence and, in this case, *inorganic*
rebirth. However, deathwishing and repudiation of the organic body—
insofar as they apply to this life and not to some afterlife or spiritual
plane—adopt a kind of psychotic and fatal reasoning, only to be
haunted by the very parts of the organic world they fail to register. For,
of course, the scientist only apparently usurps motherhood with the
extrauterine development of the robot-child; his subsequent immersion
in a sea of data is implicitly a symbolic return to the first inner space,
the womb, much as the fantasy of being eaten by machines evokes
fantasies of being eaten or destroyed by the mother.[19]

Melanie Klein's descriptions of the pre-Oedipal fantasies of infants
and very young children, fantasies that largely take the interior of
the mother's body as their mise-en-scène, bear a striking resemblance
to such imagery of immersion. Klein's model also explains fantasies of

[163]

aggression within what amounts to intrauterine space. The mother's breast, split into good and bad part-objects, is also to be found (in a sort of strange loop) in its interior; in the oral-sadistic phase, the bad breast must be pierced and punished, only to be restored and made whole again in the depressive phase, echoing cannibalistic fantasies of the breaking up and then resurrection of the body.[20]

This fantasy of being eaten by machines that, in some confused or unspecified way, are part of the natural world is graphically visualized in the controversial (and hence widely censored) industrial music video "Happiness in Slavery" by Nine Inch Nails. According to the publicity releases, a man—played by Bob Flanagan, a performance artist with cystic fibrosis whose larger subject is the intermingling of illness, pain, and sadomasochistic pleasure[21]—is shown "submitting to ritualized sadomasochistic relationships with devouring machines"; his naked body is pierced by pinchers and grinders, put into "some kind of disposal system" and, as the *Hollywood Reporter* put it, "ground into meat." Oddly enough, this strange kind of bachelor machine is "servicing the man's desires" (evidently for castration, penetration, death, and complete fragmentation). The result—in which shots of "blood and semen mix[ing] with oil" are juxtaposed with "quick cut closeups of gears grinding flesh intercut with smooth sensual moves [that] convey the sensuousness of this encounter"—appears not only to mix machine and human fluid but to nourish the natural world, which appears to consist largely of writhing worms (or, at another level, geometrically multiplying castration symbols).

[164]

Still from Nine Inch Nails/Bob Flanagan,
Happiness in Slavery.
Directed by Jan Reiss (1992).

In spite of most waking experience to the contrary, the ecstatic affect and the symbiotic relation of nature and machine are not articulated—they are just there. This is not a proposition about reality, but a fantasy that reaches back into experiences in infancy. In "Happiness in Slavery," the posture of the devouring machine leaning over the reclining, restrained body of the man to take a bite is a reminder of the archaic mother and the wish for self-annihilation to which she is ultimately linked. In this case, the oceanic feeling and ecstatic transcendence of the body occur by means of pain—and in a way that Flanagan's installation suggests is specific to this culture and its denial of illness and death.

Second Skins

In this reversible logic, in which subject and object are not clearly differentiated, rather than being eaten one can try to *become the other* by "getting into someone else's skin." That is, the cannibalistic fantasy of introjection has a counterpart in the reverse gesture, that of covering oneself with the other as a means of self-transformation. Entering this skin envelope also suggests that where the space is not already void, one is scooping out or evacuating the other, either from without or within. When an organic body as host of identity and subjectivity, rather than a clone or replicant or virtual avatar, is at stake, the struggle or sovereignity over the skin can be aggressive and deadly.

In some specific historical societies where ritual has involved human sacrifice, the "skin ego," or envelope of identity and self, could literally be transferred by flaying a human victim and wearing his or her skin. In the account of Diego Durán describing Aztec ritual, for instance, prisoners who were made to impersonate gods were then flayed: "Other men donned the skins immediately and then took the names of the gods who had been impersonated. Over the skins they wore the garments and insignia of the same divinities, each man bearing the name of the god and considering himself divine."[22] Tzvetan Todorov, who cites this account, contrasts such Aztec sacrifice of victims who are socially very much like themselves with the societies of massacre associated with the conquistadors, whose cruelty in exercising their own acquisitive and expansionist aims grew in proportion to their perception of the difference and distance of their victims from their own identities.

Literalized skin symbolism has a sad counterpart today in the flaying and facial mutilations that have been reported in the states of the former Yugoslavia, mutilation inflicted on people who are only partly other, who belong to essentially the same language and ethnicity but to different religions and cultures. The desire to literally cut away or obliterate the outer identity or "skin ego" of an other—not to mention destroying that close-other from within via hunger, rape, and torture—owes something both to sacrificial rites and a society of massacre.

When political boundaries fall apart, ego and identity are also threatened with fragmentation, and they must be radically fortified or surrender to dissolution or transformation. Even for an individual in the relative calm of the post–Cold War United States, an ordinary skin may no longer be enough to contain the ego or to protect bodily fluids from escaping or pollution and irritants from the outside world from entering. Consider the announcement of a new product, a transparent SmartSkin™ or "ultra-high-molecular-weight cellulose polymer" which is permanently electrically charged so that it firmly binds to the skin surface, now made slick, smooth, and, as a result, youthful in appearance. When used with BETAMAX CAROTENE+™, a melanin layer (or "sun tan" without the sun) is produced, in effect marrying body- and techno-chemicals into what is literally a second, fortified skin.[23]

Skin egos are usually less literal, though, as conceptualized by Didier Anzieu, some sensuous element or other envelopes the body—for instance, the muscular skin earned through weight training (also accompanied by a preferably sunless tan) or a symbolic skin applied via tattooing or writing. The ultimate second skin, though, is electronic, as in the data suit, helmet, and gloves of virtual reality. Under an electronic skin one can adopt virtually any persona and experience a written world of images and symbols as *if* it were immediate experience. Indeed, it is as if the body were immersed in unframed symbols themselves, without need for distance or reference.[24]

Immediacy and Ubiquity

Such introjecting and enveloping responses do depend on a sense of oneness and presence (what the Lacanian tradition calls the *imaginary*) but in a way in which "the distance necessary to symbolic functioning seems to be lacking."[25] Oral logic can be as archaic and *immediate* as an infant at the breast, or as immersive as the fetus in the womb. However, the expressed belief that virtual reality provides an unmediated or "post-symbolic" experience of externalized mind is an illusion fostered and supported by an oral logic of incorporation. In the state of immersion, it seems that one doesn't symbolize flying, one *does* it, just as virtual objects are, albeit only via an electronic skin. However, this illusion is possible only because the second skin (or "interface") that mediates the virtual world also masks the apparatus of that mediation. This masking allows the referential to appear to collapse into the symbolic field: to utter the symbol for an action is to perform the act itself—the classic definition of a performative or declarative speech act. (Similarly, Walter Benjamin noted that the film image is the only place in which the tech-nological apparatus is invisible, but because it is carefully organized to be out of frame.) In both cases, mechanical and electronic, this absence of the apparatus from awareness furthers psychic regression—on one hand, in the form of disavowal in relation to the classical fiction film, and on the other, as performatives licensed by various strategies of negation I shall discuss in the third section of this essay.

While "cannibalistic fantasy" may have its prototype in infancy, it is far from restricted to the past or to infantile or regressive aspects of life. One could call "eating"/"being eaten" and "enveloping"/"being envel-oped" both deep metaphors that pervade the "most advanced" cultures and the "highest" art forms. Perhaps not surprisingly, the imagery of piercing and engulfment is ubiquitous in the technological realms of laboratory-created immersive virtual worlds, as well as in high-tech war. But even certain philosophies could be lambasted by Jean-Paul Sartre for introjecting the world into the "rancid marinade of Mind." Sartre's disgust with "knowing as a kind of eating"[26] is a lucid polemic against oral logic which has much in common with Brecht's attack on the "culinary" aspects of illusionistic theater.[27]

Although oral logic is conceived as a stage of development in the infant and child before the development of language, it evidently co-exists with the logics of other stages of development into adulthood. It is never entirely abandoned and may even come to dominate an infor-

[167]

mation society. Furthermore, oral logic is hardly restricted to the thematics of food, just as food as a substance is invested with the imaginary, the symbolic, and the real, as well as participating in oral logic. Considering that food itself is the liminal organic substance at the boundary between life and death, it is also the symbolic medium par excellence. A particular cooking process not only transforms nature into culture (as elaborated in Claude Lévi-Strauss's work in structural anthropology), but it also offers the means of exchange or communion between the body, the world, and other human beings, and defines a culture per se in its specificity. There can be wide differences in the perceived immediacy or degree of mediation of the body in relation to food, from the fully enculturated eating of an organized meal with utensils to feeding at the breast (or inversely, the imaginary of being devoured by the mother) to imagining oneself inside the mother's body or immersed in oceanic oneness. This range is comparable to the different degrees of convergence of self and other, from symbolizing the other to "interfacing" with it, or wearing brain probes and "jacking in" or being "immersed" in a digital sea. (Note the implicit female gender of the space of union, suggesting that little has changed in that regard from the time of Freud to the age of information. Indeed, cyberspace is a largely male domain where gender constructs under critique in other spheres of contemporary society return with a vengeance.) Thus, subjectivity includes processes of incorporation, identification, and symbolization, and oral logic is a constant part of that range of subjectivization.

However, when fragmentation and fortification of the ego become strongly thematized, it suggests a situation of cultural distress. The contemporary prevalence of the imagery of horror and disintegration—fragmented, dismembered, or, for that matter, mismatched, multiple, or decaying bodies and lost parts (namely, the cannibalistic fantasy of the body treated as food)—suggests that something fundamental is "eating" our culture. Perhaps because we live in a situation of epochal cultural change, envelopes containing cultural identity—the body image and skin ego—seem to have been torn beyond repair. Provided one does not accept dissolution and transformation into something other than what one was before, a body whole and entire must somehow be restored. But how can the body be resurrected when it is so loathed?

We are in a strange situation when the desire for fusion and wholeness presupposes, at least in representation, the repudiation or disavowal of the body and the negation of food itself.

As the dark mirrors of Cold War identities shatter and the power of the Face (which is implicitly white) wanes,[28] we stand at the beginning of an epochal change—for which, of course, we need (non)bodies to match. Our imagery is no longer one of confrontation with the other at well-defined borders, but a confusing zone of shifting culinary and symbolic boundaries. That a cultural boundary has been crossed is often not signaled by lifted gates from without but by nausea from within, when, perhaps unbeknownst, cicadic or other bodily rhythms have been disturbed or the local flora of microbes are wiped out or displaced.

The answer of electronic culture to a vision of confusion and waste is largely one of "purification" or disembodiment. However, this option may be of limited value, for how can the response of culinary and corporeal negation afford to be more than a minor and transitional phase when food confronts cultural change? Food is at once a symbolic system and organic fuel; thus there are limits to *nonfood*—we humans must eat or perish. However, a tacit assumption of the contemporary technological fantasies to be discussed later is that we perish *because* we eat. The desire for an evolutionary transformation of the human has shifted focus from the preparation for the journey into "outer space" from a dying planet to the virtual "inner" space of the computer.

II. Failed Food Ideologies and the Real

The failure of a food discourse can be read in terms of survival of a population; the body itself is the surface written and sculpted from within in terms of well-being or in terms of eating disorders, disease, and death. Which is to say, food ideologies and the symbolic order have an ineluctable organic limit that might be thought of as the intervention of the *real*. The consequences of the sociopolitical and ideological dimension of food can be seen in a malnourishment for some and an overabundance for others that cannot be blamed on natural forces such as the weather or the fertility of the soil. To an outside observer, the most striking American culinary metaphors might be: too little, too much—on the one hand, lack of food for significant populations of children, the working poor, and the homeless for at least part of each month and poor-quality food for the rest of it.[29] On the other hand, the significant number of "overweight" Americans suggests a perverse situation of unwanted abundance. In Berkeley there is a good-hearted but absurd- sounding organization for passing along the calories of dieters to the homeless; yet it would take just such an actual transfer of caloric capital on a massive scale to have an effect on national health. Considering this caloric imbalance, the existing American techno— fast-food system has discredited its democratic and utopian potential of feeding everyone cheaply and well.

Fat: The All-Too-Visible Flesh

However ubiquitous cannibalistic fantasies in our culture, the American public was not prepared to regard the images from Somalia of taut skin over fleshless skeletons (with eyes that could return our gaze) with any-thing but horror. The images on American television at dinnertime conveyed not just human suffering in a situation of social and moral collapse: it confronted Americans with what amounts to our own ideal image taken to extreme and inverted in value. (The story of a reporter drinking diet soda amidst the starving is already legendary.) The fam-ished bodies of Somalians were seemingly sucked dry, excavated by an immense malignant force in a grotesque and tragic exaggeration of a fleshlessness so often idealized (but so seldom achieved) in America. When we find emaciation in America, it can well be malnutrition, but it can also be a fashionable result of willpower *and/or* the effects of an eating disorder (such as bulimic purging). The American body may often be hungry, but it achieves most cultural visibility in the United States not in terms of fleshlessness but as fat.

[170]

While some critics have labeled the press coverage in Somalia "disas-ter pornography," the gaze of American viewers is certainly one not of guilty pleasure but of guilty repulsion. Advisers are reported to have counseled American soldiers embarking with foreboding for Somalia that it is normal to feel sick on encountering victims of famine.[30] Considering the relative lack of discourse on hunger in the United

States, it is ironic how much public support there was for what amounted to a militant response to the devouring land of Somalia, until the sight of a mutilated American corpse appeared in the news. Whatever the real political agenda that welcomed an American military presence in Somalia, it was television coverage of starving Somalians that provided the occasion, and the horror of American bodies in pieces that set the limit for American involvement. Even intact American flesh—when not covered by hypertrophied musculature or second skin—can be an object of ambivalence and even moral revulsion.

In America "fat" is a stigma and the sign of the self-indulgent behavior of someone who has "let him- or herself go." Yet the fat body has an intimate and causal link not only with a poor-quality, fast-food diet (shared even by an already labeled fast-food President Clinton, counseled to freshness by celebrity chefs and schoolchildren), but also with the life styles of an information society. Consider, for example, the link between a massive increase in pizza deliveries in Washington and the high-tech planning for virtual war in the Gulf. But the link is even more direct: recent research shows that just watching the tube can be fattening.[31] Perhaps, as Michael Sorkin suggests, writing on the future of design, Walter Hudson, the 1,200-pound Guinness record holder for body weight who died Christmas Eve 1991, "was the ideal citizen of the electronic city,"

> not for his bulk but for his immobility. Surrounding his bed was a kind of minimum survival setup—refrigerator and toilet, computer, telephone and television—a personal pod. The system had Hudson exactly where it wanted him, fully wired in, fixed in location, and fully available to both receive signals and to provide a stream of negotiable images....we are all at risk of becoming so many Walter Hudsons, well-wired lumps of proto-plasm, free to enjoy our virtual pleasures, mind-moving and disembodied, unable to get out of bed.[32]

By embracing the results of today's sedentary life style and a fast, high-fat diet tongue-in-cheek, Sorkin's ironic image of the future confronts the hypocrisy of a society that does one thing (or actually nothing but small motor movements) and values another (the body as perfect human-machine). Such all-too-visible flesh puts what some consider a crisis of (in)visibility in the age of computers in another light. It is true that without visualizations for the monitor screen or phantom journeys "inside" the computer via virtual reality and science fiction, computer events offer little for human eyes or ears to perceive. But to see the problem of information as invisibility alone is to disavow the intimate connections between the virtual electronic story world (what happens inside computer systems and networks) and the very visible and increasingly repulsive and/or wasted world of the organic body outside the screen.

While fast-food restaurants have begun to respond to health criticisms by lowering fat content—without much commercial success—a recent PBS documentary on the industry, *Fast Food Women* (1992), suggests the social cost of cheap, albeit addictive, food: repetitive work that is not only unsatisfying to perform but that does not even return a living wage. In fact, employment in the fast-food industry has become the emblem for a postindustrial trend toward the deskilling of the labor force and an expansion of a low-wage service sector; at the same time, there is a counterdevelopment of a highly educated and skilled information elite of what Robert Reich calls "symbolic analysts," in his *The Work of Nations: Preparing Ourselves for 21st-Century Capitalism.* Cheap, fast food seems to contribute the social disparities its accessibility appears to heal. The addictive power of drug-food ingredients also enhances disengagement from unpleasant surroundings, not unlike other phenomena of electronic culture, from virtual personas to smart drugs.

Fresh Failures: Tourisma and Frankenfood

[172]

Perhaps those very symbolic analysts who eat fast food when working can best afford to be "foodies" during leisure hours. For while the once utopian food counterdiscourse of *freshness* in America still retains many of the populist and progressive aspects it had in the 1960s (such as engagement in practical issues of healthful ingredients, pesticide use, and small farm production), in the 1980s it evolved into an elite restaurant culture of "foodies" grazing on tiny portions. While the values of *freshness* may harken back to a preindustrial food system (indeed, what French farmers with pitchforks are now fighting to preserve in the face of McDonald's), restaurant food art has actually come to rely on expensive ingredients delivered by air and van from small or boutique farms. There are many signals in current food discourse that the "nexus of haute cuisine and counterculture" which opened in 1972 was over in 1993; some have claimed that we are at the end of a cycle when "culture's culinary frame of reference has simply grown wider and sturdier and more self assured"[33] but, as other style-section headlines exclaim, "The party's over."[34] Perhaps in the 1990s it is not possible for those at the feast to ignore the others who are outside dumpster diving.

The social failure of the "fresh" is a sad one that includes food critics as well, who rarely address what a homeless person eats for breakfast or what is being offered in school lunches today. (Certainly, one of the fundamentals of human intelligence, not to mention social justice, is having enough food to eat of sufficient quantity and quality.) In what appears an oblique response to a more general economic decline, high-art restaurant culture seems to have backed away from eclectic postmodern concoctions in favor of comfort food that doesn't draw attention to itself, a zero-degree of culinary writing. The results are simulations of food like Mom would have made had she cooked a simple ethnic cuisine without the schmalz.

Still from Ken Feingold,
Un chien délicieux (1991), 19 min.

But one might ask whether the border-crossing mixtures of ingredients for which fresh food art was known were not actually a form of culinary tourism rather than multiculturalism. That is, to mix the signs of difference—particular herbs, spices, and other ingredients, of say "Mexico" with "France"—is not to cross actual cultural boundaries. Even the most adventurous "food art" maintains certain cultural limits, as Ken Feingold's 1991 video *Un chien délicieux* demonstrates. In 1986 Feingold recorded an interview with a Burmese man from the Golden Triangle of northern Thailand. Later he added a voiceover in which the man is reported to have reminisced about Paris and his friend André Breton of the Surrealists, in what amounts to a recipe for a "Surrealist cookbook." As "Lo Me Akha" was preparing to leave Paris, "Breton" (seen in a well-known photo) promised him a meal of his choice. Lo picked dog, an Asian delicacy—a taboo choice Breton was not eager to prepare. Feingold's video includes documentation from 1986, framed tongue-in-cheek as a cooking demonstration that suggests the sacred limits of what many in the West consider good to eat and, perhaps, the parochialism of postmodern food eclecticism, as well as the dubious nature of reportage—especially across cultures. Revulsion and what is more than jokingly called Montezuma's Revenge are the surest indications that one is crossing the boundary beyond tourism to, if one would have it, the Other. For culinary cultures are not cruise-style smorgasbords; their borders are guarded by microbes and nausea. Indeed, it was just such an abject moment during the Bush presidency in 1991 which, in retrospect, marked not only its decline, but an end to post–Cold War euphoria—the sight of the president vomiting and collapsing into the arms of his Japanese host.

There is also a fundamental threat to the ethos of freshness in its very nearness to the *raw*—the growing mistrust of the natural. Technological advances have produced "fresh" food that virtually can't decay, via irradiation and genetic engineering. Nuking microbes to preserve food suggests a state of undeath rather than freshness. Meanwhile, the genetic material that allows vegetables to decay can be inverted, producing a kind of "zipper" effect against rotting. Such "frankenfood" has been perceived as more than mere chemical adulteration by advocates of

[174]

freshness: "Don't put trout genes in my tomatoes!" exclaimed a San Francisco celebrity chef in disgust during a talk-radio interview. And no wonder, since by simulating freshness, genetic engineering and the production of frankenfood blurs the most fundamental distinction between industrial fast food and its culinary postindustrial counter-ideology: it may look unspoiled, but it is symbolically as well as geneti-cally contaminated. In retrospect, it appears that what is at stake in fresh cuisine is not freshness per se but a continuum of ripeness to rottenness; the fresh is a vote for the organic itself and the mortality with which it is inevitably at one. Unfortunately, such an ideology appeals largely to those with the means to hold decay at bay.

Yet even the discourse of health appears to have abandoned, in part, freshness. Now there are ex-"foodies" concerned enough with health maintenance and intelligence maintenance to have, insofar as possible, stopped eating unhealthy and, at the extreme, dangerous substances (namely food) from depleted and often polluted soils largely lacking in the trace minerals and electrolytes needed by the brain.[35]

Since food is a lived metaphor of culture itself, it should not be sur-prising that a culinary system with many analogues with the computer could emerge in the United States. After all, the television/microwave and food/word processing have been image supports of the "fast" and "fresh," the two great food ideologies of the postwar era. But the com-puter and cyberspace have come to be reference points for what amounts to a postculinary discourse. Perhaps, as opposed to Alice's "very small cake, on which the words 'EAT ME' were beautifully marked in currants," an electronic culture virtually confronts us with the direc-tive, "DO NOT EAT." For what cyborgs eat (and what evidently incorpo-rates people into cyborgs) negates the very idea of food as mediation between the organic body and the natural world.

III. Post-Culinary Defense Mechanisms

The *negation* of the organic body, its nourishment, and all that the body stands for can occur in many different cultural fields and adopt many different means—for instance, forms of psychic defense such as *repudia-*

tion, denial, or disavowal.[36] Furthermore, the body–machine relationship can be inverted, much as the body may (at least in fantasy) be turned inside out in surrender to the collapsing boundaries between the symbolic and meaninglessness, as in Julia Kristeva's concept of *abjection.*

These modes of negation are at work in contemporary phenomena such as cyberpunk fiction, simulated food, virtual reality, smart drugs, and finally, "excretory" art. However, the texts to be discussed here represent or present these cultural phenomena with varying degrees of distance to what they portray—from critique, irony or cynicism, and positive engagement to a perverse kind of heroic idealism. What we are dealing with here is not the psychic defense itself, but its use in a *symbolic* coming-to-terms with cultural distress. Here is a playful realm of the subjunctive in which the fundamental mode is disavowal, or split belief: I *know* it's just a story (a fake, an optical illusion, a fetish, a performance), but *nevertheless....* It is primarily the literal, serious manifestations of such phenomena that can be dangerous, for good or ill.

Purification Strategies: Repudiation and Cyberpunk

Repudiation is the "rejection of an idea which emanates from external reality rather than from the id. It is a failure to register an impression, involving a rejection of or detachment from a piece of reality."[37] What certain cyberpunk fantasies fail to register is the organic body itself. If, according to Allucquére Roseanne Stone, cyberspace is "a physically inhabitable, electronically generated alternate reality, entered by means of direct links to the brain—that is, it is inhabited by refigured human 'persons' separated from their physical bodies, which are parked in 'normal' space,"[38] the next question might be, how does an organic body "park"? The answer of William Gibson's sci-fi novel *Neuromancer* is to submit the body to pseudo-death in the coffins and loft-niches of the desolate landscape of Chiba. While the surrogate Case travels cyberspace ("Unthinking complexity. Lines of light ranged in the nonspace of the mind, clusters and constellations of data. Like city lights receding"), his organic body is evidently in a state of suspended animation, neurally sustained by a fantastic pharmacopeia.[39] The junction between the human brain and the computer ("a graphic representation of data abstracted from the bank of every computer in the human system"—

note slippage between computer/human) seems to consist of electronic impulses between implanted chips and brain chemicals enhanced by drugs.[40] In terms of oral logic, machine penetration of the brain allows the fusion of electronic-human chemicals, which in turn allow the virtual traveler to be enveloped in the electronic skin of cyberspace—at the cost of leaving the meat behind.

But the flesh is left gladly. The desire to repudiate the body that pervades the novel—a desire that, of course, can't succeed in reality without a fall into a completely psychotic state—is marked by revulsion at the very thought of "meat." Even the description of sex reads like an SCSI port docking. As for food, in *Neuromancer* there is a restaurant called the Vingtième Siècle where steak is served. But Case isn't hungry, despite Molly's cry that "They gotta raise a whole animal for years and then they kill it. This isn't vat stuff." Why? Because his brain was "deep-fried" and the "aftermath of the betaphenethylamine made [the steak] taste like iodine."[41] (Corpulency will never be a problem for this hero.) Ultimately, (im)mortality and the (in)capacity to love become the principal issues in the novel, though not in the way they might in Chiba but, rather, in the ethers of cyberspace—suggesting that for virtual bodies (and, by extension, for cyborgs) the limits of organic life are not evaded but merely displaced.

Beyond Operation Margarine: Simulation and Olestra

Denial is a way of negating the corpulency-prone, mortal body (that is, "the piece of reality") at the *symbolic* level. Grosz describes it thus: "By simply adding a 'no,' to the affirmation, negation allows a conscious registration of the repressed content and avoids censorship. It is a very economical mode of psychical defense, accepting unconscious contents on the condition that they are denied."[42] Of course, *denial* as Roland Barthes identified it in his well-known "Operation Margarine" can also work at a far more conscious level of willful public and self-delusion. In *Mythologies*, Barthes described the process of denial at work when confronting unpleasant cuisine like artificial fat—at a time when the "natural" was preferred. (Today, of course, a preference for the artificial prevails among many initiates of an information society.) First, the fact is affirmed (that is, *yes*, it is margarine), then denied (*but it* is, in effect, butter) in an act which inoculates the discourse against the artificial, after which the fact may be ignored without embarrassment.

On the other hand, Olestra, a synthetic fat product patented by Procter and Gamble, does Barthes's margarine one better: "It [sucrose polyester] retains the culinary and textural qualities of the fat, but in a form that the body is unable to digest. Result: a fat that passes straight through the body." As a result, one needn't even try to negate or repudiate the body—its cravings have no consequences for health or mortality. One can "let oneself go" or "carry on" as before and nonetheless be pure, because Olestra is not merely artificial (like margarine, which remains a fat, with all its consequences for the body): it is the simulacrum of food, a solution to the fat accumulating on this side of the television screen and computer monitor, one that requires no change in lifestyle to purge the flesh. It is this process that makes junk food— itself already a concoction of artificial food ingredients plus sugar and fat—subjunctive or contrary-to-fact food, that is, "Junk Food That's Lean and Healthy?"[43] (Note that consumer groups *and* the National Renderers Association have protested Procter & Gamble's plans to market the product.)

Interestingly enough, this "fat-free fat," another guise of nonfood, is the opening metaphor for artificiality per se in Benjamin Woolley's introduction to *Virtual Worlds*.[44] For Woolley, this culinary bypass operation raises the issue of what remains real in an increasingly artificial world. Unfortunately, his argument resorts to denial in another mode: Woolley eventually finds reality itself to reside not in a visible material and physical world, but "in the formal, abstract domain revealed by mathematics and computation."[45] The capacity to model mathematically or *simulate* what is otherwise beyond human perception is thus more real than whatever passes for reality itself.[46] So, on one hand, there is a "food" that treats the body like a cyborgian steel conduit but allows us all the incomparable pleasures of junk food; on the other, there is Woolley's implicit rejection of the visible and manifest (here the "flesh" and its problematic nourishment) in favor of truth in pure mathematics—an act of denial. Neither Olestra nor math ultimately evades the effects of the real.

Virtual Reality and Telepresencing:
The Body Disavowed

In virtual reality, the "meat" body is not "parked" but rather mapped onto one or more (possibly shared) virtual bodies; at the same time, the organic body is purified by seeming to be out of the frame, hidden from view for the virtual traveler, in favor of a cartoonlike graphic world. It is as if Walter Hudson were able to shrink down to whatever shape he desired and enter his TV, transformed into fully rendered, life-size 3-D space filled with virtual objects (and possibly other virtual personas) with which he could interact. To "enter" or "stick your head into" such an immersive artificial world with a technological second skin—or helmet and gloves—simultaneously blinds the virtual traveler to the world and obscures from view his/her organic body *and* the machine apparatus that sustains the virtual world. Meanwhile, an organic finger merely points inside the dataglove and the surrogate body flies at great speed through an artificial world, promoting an impression of disembodied superpower and almost omnipotent thought in a persona that is freely chosen, not contingent, and limited by flesh.

It seems the apparatus of virtual reality could solve the problem of the organic body, at least temporarily, by *hiding* it. Yet, the organic body as problem has not been eluded: it has only been made momentarily invisible to the user. According to Michael Naimark's "Nutrition" segment of the tongue-in-cheek video documentation, *Virtuality, Inc.*, virtual reality even has application as a diet tool (and, I might add, in a way not unlike Olestra): while a helmet with television "eyephones" blinds

Still from Michael Naimark and Students of the San Francisco Art Institute, *Virtuality, Inc.* (1990).

the woman immersed in virtual reality to what we as spectators see are actually nutritious crackers, she munches with moans of pleasure on virtual cherry pie.[47]

The seduction and playfulness of virtual reality are based on this very disparity between organic and virtual bodies—its power to erase the organic from awareness, if only partly and just for a while. To the degree that the duality of worlds (a reversal of the everyday situation of mental invisibility and physical visibility) remains conscious to the one immersed in a world in which s/he possesses superpowers, the situation is one of *disavowal* or split-belief ("I know it's just a computer-generated display, but nevertheless...").[48]

However, once one switches one's viewpoint from the internal to the external, from the virtual to the organic world, virtual reality takes on a wholly different guise. To the observer on the outside, the improbable gustatory moaning of the eater of the virtual cherry pie resembles a regression to infancy; similarly, the flailing motions of "flying" (or other virtual locomotion) suggest an actual situation of helplessness and vulnerability in physical space.[49] Indeed, immersion in the artificial realms of information presupposes not only an electronic skin but also a womblike fortress of safety from the physical world *before* the user can enjoy apparent invulnerability.

This link and disparity between the two worlds can serve virtual play. But the two worlds can also be electronically linked to cause real effects in other distant (or microscopic) parts of the physical world via *telepresence*, as in, for instance, long-distance robotic brain surgery or precision bombing. Organic bodies in the referent world (that is, those without access to the virtual system who are to be operated on, bombed, et cetera) need to be prone, anesthetized, or otherwise without power over the controls in order to be vulnerable to the actual remote operator of any robotic agent. It is as if persons with material bodies were confronted with phantoms; or, as if they were put into a story as characters within which other characters are surrogates of the author and enjoy his or her special privileges and ultimate invulnerability. The danger in the electronic divide between symbolic analysts encapsulated in global cyberspace and those outside is a kind of willful blindness that supports that inequitable distribution of power (not to mention calories and culinary capital) that is, ultimately, a negation of whatever social contract might govern conduct in material space.

Furthermore, responsibility for the organic consequences of remote action are all the easier to deny: consider the virtual conduct of the Gulf War and the many opportunities it offered for denying the connection between war and human suffering. Yet, even for relatively invulnerable warriors beyond the phantoms and under the technological superskin, the temporarily invisible flesh that suffers hunger and that needs to go to the bathroom is still there, its demands merely deferred.

Smart Fetishism: Do Not Eat

Smart drinks and drugs are the ultimate fetishes for initiation into the cyborgian condition. Like tiny introjected phalluses (that is, undecidably organic/electronic and thus both), they offer a kind of magical thinking, the promise of human transcendence, with the alibi of science. This imaginary ideal nourishment is a technofood reduced to its byte-sized chemical constituents like decontextualized data for intake into a brain conceptualized very much like a computer.

At present, smart concoctions actually consist largely of vitamins and drugs developed for the treatment of Alzheimer's, Parkinson's, AIDS, cancer, and other diseases. This suggests that its deepest rationale is fear and its modus operandi the preemptive strike. Some people who once preferred baby vegetables now eat choline and other amino acids, minerals, herbs like ginkgo biloba and ginseng, drugs like piracetam, Deprenyl vinpocetine, aniracetam, pramiracetam, oxiracetam, pyroglutamate, and other cognitive enhancers such as AlC, caffeine, Lucidril, Al721, DHEA, SMAE, Ferovital, Hyderine, Idebenone, Phenytoin, Propranology Hydrochloride, thyroid hormone, vasopressin (that is, pituitary gland hormone), vincamine, vitamins B, C, and E, Xanthinol nicotinate, essential fatty acids, selenium, L-Dopa, RNA, human growth hormone, and the neurotransmitters-noepinephrine, NE PRL-8-53, and ACTH4-10.[50] Such a chemical litany also serves the deniability factor: what is smart isn't really food—but, then again, it is.

In addition to ex-"foodies," people in cyberpunk circles and techno-music clubs have made the smart (and the psychedelic) the beverage of choice, in lieu of snacks, alcohol, and Coke. The odd result is a mix of the discourses of health, space exploration, and junk food. No wonder, then, smart drugs enjoy an at best quasilegality (but then so should barbecue potato chips). In youth culture "raves" and technomusic clubs, smart drinks and drugs (or "nootropics," which provide "desirable qualities of cerebral stimulation without the negative side effects of ordinary psychoactive drugs"[51]), enhance the effects of psychotropics like XTC and MMDA, extremely fast-beat technomusic, and the psychedelia of image and light shows, whistles, and special glasses to create a communal sense of oneness and high energy in a self-sufficient and completely engrossing present tense. While the psychotropics are on the Food and Drug Administration's (FDA) Schedule One (that is, they are illegal), the nootropics or smart drugs have had a hazy legality when ordered for personal use without prescription from England or Switzerland under a hard-won circumvention of FDA rules via AIDS activism. However, the FDA's recent raids on vitamin dealers have put even smart drinks largely composed of vitamins on notice. Perhaps, as with rock culture, this semioutlaw status serves the ethos of "smartness" as a counterdiscourse with a program for, if not social, then evolutionary change out of the human condition.

There are certain recurring features in the very limited literature on smart drinks and drugs in how-to books, manifestos, and ads in *Mondo 2000*: smart nonfood tastes bad—medicinal, in fact; smart drugs are better than nature, once one achieves the right "fit" between brain and chemicals; and they improve performance in mental tasks. To at least one countercultural theorist, Terence McKenna, smart drugs, insofar as they are psychotropic, are in fact *Food of the Gods*, at once archaic and posthistorical tools toward the next phase of human evolution toward colonizing the stars.

Bad-Tasting Medicine

[182]

Once we have entered a realm of negation and nonfood, we have left behind as cultural values the voluptuous effects of food on the "gastronome's body" of a Brillat-Savarin or the sense of "well-being" Roland Barthes describes as *cenesthesia*—the total sensation of the inner body or bowels.[52] For if food is the manna of fullness and pleasure, nonfood is bad-tasting medicine that—precisely because it is disgusting—can be eaten with pleasure, much like the ecstatic response to devouring machines.

While an April 1991 cover story on vitamins in *Time* stressed their supplemental character and claimed that "real food is here to stay" (if nothing else, for reasons of obscure nutrients—for example, phenols, flaveins, and lutein—and for "hunger and the savoring of good food"[53]), other prophets of foodceuticals as longevity enhancers described with particular relish one of their concoctions (arginine and cofactors) that not only didn't taste good, it smelled like "dog vomit."[54] Pills as electronic metaphors for firing synapses may not taste very good. Evidently, savor and taste are not the primary issue when "smartness" or health are at stake.

The underlying image may be a mixture of medicine and the future-food of aerospace and astronauts, but at stake is *not* the legendary unpalatability of K-rations or MREs. Rather, ingesting this space-age concoction is part of the magic of preparing for a future in which more and more demands will be made on mental performance. While—or even because—smart drugs are not so very good to eat, they may be literally good to think. Smart-drug discourse vacillates between medicine that is "good for you," an ascetic or virtuously masochistic negative desire to transcend the body, and a cerebral high that invokes a body image of wholeness and perfection at one with the future.

Better Than Nature

Smart discourse how-tos are largely about the right dose and the proper fit between chemicals and the brain. Of course, "fitting" the various entities, reality statuses, and modes of electronic culture together is the general practical problem a global information society must solve in order to come into being; composing cyborgs is but an especially difficult instance of it. Note that the cyborgian direction of fit between organic and electronic is heavily weighted toward the latter.

Smart drug "fit" is not based on existing "natural" quantities— neurochemicals are too costly for the body to make in beneficial amounts.[55] However, according to Terence McKenna, nature has offered psychoactive drugs, which are not merely smart but, he claims, have spurred human mental evolution, in abundance. In *Food of the Gods*, McKenna explains, "My contention is that mutation-causing, psychoactive chemical compounds in the early human diet directly influenced the rapid reorganization of the brain's information-processing capacities. Alkaloids in plants, specifically the hallucinogenic compounds such as psilocybin, dimethyltrypta-mine (DMT), and harmaline, could be the chemical factors in the proto- human diet that catalyzed the emergence of human self-reflection."[56] McKenna views the fifteen thousand years of cultural history between the archaic period and the present as "Paradise Lost," a dark age of ego-imbalance to be abandoned, along with "the monkey body and tribal group," in favor of "star flight, virtual-reality technologies, and a revivified shamanism."[57] Again, the archaic and the electronic are united.

Smart Performances

It is informative to consider what "smart" means in the context of this drug discourse, where "learning" is defined as "a change in neural function as a consequence of experience."[58] In descriptions of the drugs, smartness implies more effective neurotransmissions; however, in drug testimonials by users, "smart" is not described in terms of higher cognitive processes but as the ability to retrieve trivial or obscure information in the context of school or work. This information recall is prized largely for its exchange value or as evidence of performative ability and instrumental reasoning capacity: a secretary given a raise by her boss to buy smart drugs becomes "more alert, and intelligent acting and she smiles more. She is overall a much better employee." A student is enabled to become a math major and get a job in Silicon Valley. A graphic artist is able to work all night and present her work the next day with a smile. A father in his forties is given Hydergine by his son, and to the son's amazement, the father recalls "family vacations, picnics and holidays" that happened in his twenties."(!)[59] (Note how often "smartness" is something desired of someone else.)

Smart drugs also reportedly rejuvenate sexual performance a good twenty years.[60] Yet often, however facetiously, descriptions of sexual and keyboard activity are mutually substitutable or are metaphorically intertwined: "It started so innocently: just a snort of vasopressin before sex, or before getting down with your keyboard."[61] So, smart drugs may enhance cognition and sex (or they may not), but many of the motivations the discourse implicitly suggests for taking them imply a situation of stress and fear for the future, plus wishful thinking. My reading of the connotations of smart drugs extends beyond the personal utopian quest for health and longevity to include loss of faith in our ability to survive a toxic natural and social world without medicinal help, as well as guilt and despair with the arrangement of our social-communal world. The half-secret and intensely shared present tense of a "rave" is a substitution for such communality. Capsules of "information" are at once a kind of sympathetic magic that allows the body to converge with computers and apotropaic magic that holds at bay all sorts of plagues now loose in the world.

Contamination Strategies:
Excremental Art and Cyborg Initiation

If nonfood is a cuisine of lack *or* waste, nothing or excess to be received with loathing, then rather than being taken in, nonfood may be something that is spit out. And rather than purifying the body into electronic wholeness, the negation of the organic may wreak a transformation via waste and defilement, in which bodies are violently implicated, torn open, their inner linings exposed to the world, allowing bodily fluids and food waste to smear the boundary between inside and outside, and between self and other.

Such deliberate violation of the surface of the body or skin affects the symbolic order as well by undermining the boundary which produces recognition, identity, and meaning by separating it from a ground of meaninglessness: namely, a deliberate evocation of what Julia Kristeva has theorized as abjection. In *Powers of Horror*, she describes the primordial experience of abjection as food loathing—in her case, disgust at the skin forming on the surface of a glass of milk. Spitting the milk out establishes a limit, a not-self, but at the cost of expelling a substance, food waste, that is ambiguously self and not-self. Kristeva writes: "I expel *myself*, I spit *myself* out, I abject *myself* within the same motion through which 'I' claim to establish *myself*…it is thus that *they* see that 'I' am in the process of becoming an other at the expense of my own death."[62] Beyond food loathing, other examples of abjection include the sight of bodily fluids or corpses. She concludes, "It is thus not lack of cleanliness or health that causes abjection but what disturbs identity,

[185]

system, order. What does not respect borders, positions, rules. The in-between, the ambiguous, the composite."[63] To embrace symbolically the abject is to evoke horror and to call forth this disturbance and this death, in order to become something else, now shapeless: one identifies not with the body, but with the waste.

Excretory art is a tendency of photographic, performance, and installation art of defilement characterized by smeared foods and the simulated ooze of bodily fluids. The notion of impurity (that which is transgressive or forbidden) is central to the use of such outlandish or impolite materials by a wide range of contemporary artists from Piero Manzoni and Cindy Sherman to Mike Kelley and John Miller.[64] Certain of these works have inspired the condemnation of conservative politicians, most notoriously Andres Serrano's crucifix immersed in urine and Karen Finley's smearing of simulated feces (actually chocolate) over her body.[65] The voluptuous effects and gustatory pleasure of eating are inverted into disgust and food loathing: the gastronome himself has become the artist as homeless person, as displaced woman, as border creature, homosexual, and plague victim. Using apparent food waste and bodily fluids as media simulates exposing the inner lining of the symbolic to light, in a kind of writing on the margins that is not now clearly delineated enough to be deciphered. At stake in the arts of disgust is the symbolic itself and the generation of new subjects, tongues untied.

Judith Barry, *Imagination Dead Imagine*
(1991), 5 screen video projection,
114 x 96 x 96 inches.
Courtesy Nicole Klagsbrun.

No wonder the unerring conservative Senator Jesse Helms introduced legislation that would have prevented the National Endowment
for the Arts from funding any artworks describing or representing
sexual or excretory organs or activities.[66] This genre of expelled secrets
and exposed linings is a whipping boy for iconophobic politicians, who
mistake symbolic acts or statements for the cultural instability and decay
to which they allude. This child being beaten is a stand-in for many
others in a culture undergoing epochal change.

A recent monumental sculpture by Judith Barry takes the theme of
contamination one step further: Barry used complex digital technology—the capacity to turn images into electronic information or pixels
—to mix the images of two live models (one male and one female)
together into an undecidably gendered huge human head projected on
four sides of an eight-foot-high cube.[67] The majestic scale, geometric
purity, and luminosity of the video head make it seem sublime, like an
ancient sculptured head of Athena; yet, what we hear is the more and
more labored breathing of the "being," a mysteriously androgynous and
lifelike electronic persona or cyborg trapped within.

Barry's title, *Imagination Dead Imagine*, was borrowed from the title of
Samuel Beckett's last novel, "possibly the shortest ever published," perhaps because what it tersely describes is the end of stories, a cycle
between extremes where nothing else conceivable could happen next.
In a scene of geometric purity, light, and heat that passes into dark and
cold and back again, a male and a female suffer the cycles of light and
dark aligned in matching semicircles; their gazes never meet except
once, at the beginning. Barry's *Imagination Dead Imagine* distinguishes
itself from its Beckett namesake by blending what Beckett so carefully
segregates: "dead" takes on the connotations of whatever exists beyond
secure trajectories and boundaries—male/female, alive/dead, human/
electronic.[68] At an elegiac pace, the digital being is repeatedly anointed
with disgusting substances. In each of the eight three-dimensional video
sequences (in a shoot that required three days, five cameras, a ten-per-
son crew, and a professional special-effects technician), what appear to
be "bodily fluids"—urine, blood, feces, semen, vomit—or substances
associated with corporeal decay—bugs, worms, sand—are poured over

the head and allowed to gush or trickle down its face, sides, and back. Each substance flows down the head in sensuous, even disturbingly erotic colors and patterns, whereupon the head is "wiped" digitally clean in order to suffer the next indignity. The video cycle takes about fifteen minutes to demonstrate eight varieties of dreck; almost heroically the head suffers humiliation after humiliation, remaining impassive but for the eyelids closing to protect itself against each deluge. The only sound is the magnified sound of breathing and the noise of each anointing—the blood spatters, the crickets chirp, and the earth makes a falling sound.

Barry was also influenced by Antonin Artaud's essay, "All Writing Is Pigshit," which suggests that the stains and blotches of bodily fluids could be a kind of writing that we can't read yet, a language in pain, caught before it can come into shape. In eight untitled drawings that accompany *Imagination Dead Imagine*, Barry defiles words rather than the undecidably human/electronic. Each drawing consists of a series of words, each in a different typeface drawn in pigment on handmade paper, then rubbed and stained to the edge of legibility with dirt and insects, meal worms, soup, blood, tea, glue, beans, and vinegar. The words are chosen from the semantic range of awe and horror, such as: "exalt," "gag," "bilge," "engross," "spew," "defile," "refuse," "transgress," and "ameliorate." Together these terms build a growing sense of excess, which is amplified by the way their lines and edges are nearly obliterated, in graphic expression of ecstasy and abjection, caught in the pain of becoming.[69]

The bizarre worm sequence of *Imagination Dead Imagine* also points to a transition from one state to another: the filthy face deluged with worms that writ(h)e and crawl over the electronic head is imperfectly superimposed over a clean face with fluttering eyelids. The result simultaneously suggests being buried alive and its opposite, resurrection. So, the body in decay coexists with the body coming to life: a disgusting spectacle, like a horror film in the part we want to look away from but can't. Thus *Imagination Dead Imagine* covers the electronic with second skins of symbolic mortality.

Many historical and contemporary cultural rites throughout the world propitiate the spirits of the dead with food and libations. (For example, in Chinese rites humans may eat the delicious offerings to ancestors with gusto—in effect, incorporating them; other cultures prefer to leave the food and wine of the dead or of sacrifice to the deity to evaporate and decay.) Considering widespread practices of making offerings to inanimate spirits, it is not so strange to subject immaterial projections or the ghosts in machines to symbolic exchange with death. What are these devouring ghosts but alienated human agency, otherwise locked out of ripeness and development in time?

Implicit in the very question of what cyborgs eat is an accommodation of the human to the machine. The better question could be: How can cyborgs incarnate the human condition? How can cyborgs become meat?

Three Paradoxes
of the Information Age

Langdon Winner

The prevailing ideology of technology in our time, one that has
endured for two centuries, bears the name "progress"—the belief that
living conditions for the world's population improve through scientific
and technological advance as applied in economic development. Today
it is possible to suppose that new ways of thinking about the human
prospect might arise to challenge this threadbare faith inherited from
the Enlightenment. Although much discussion now focuses upon alternative paradigms, the idea of progress is still firmly implanted in the collective consciousness of Western industrial societies.

At a political level, the old creed expresses itself as an almost universally shared commitment to economic growth. All politicians, especially
those running for election, must swear knee-jerk allegiance to this central goal of political society. Indeed, mindless talk about growth has
become a common speech impediment, one that afflicts the ability of
citizens and politicians to imagine other ways in which technology
affects political society, making it virtually impossible to discuss the ends
that underlie received dogma, let alone any fresh approach to understanding the relationship between social policies and technical means.

The strongest doubts about conventional notions of progress now
hinge on environmental concerns. Some observers are prepared to
affirm that runaway technological change and limitless economic
expansion are a threat to life on the planet. From that standpoint
growth must be renounced because it is ultimately a destructive pursuit.
But that realization seems unlikely to attract a wide following until
unmistakable signs of ecological devastation finally destroy the popular
consensus.

[191]

There are, however, additional reasons to question society's commitment to the ideology of progress. Economic benefits are supposed to trickle down: the less wealthy segments of society may not benefit and may even be harmed by technological innovation, but they can look forward to jobs, income, and improving social conditions as technical improvements and economic prosperity reverberate. Alas, what people today see trickling down is unemployment, urban decay, rural poverty, illiteracy, drugs, dysfunctional communities, and shattered lives. It's difficult to decide whether the biosphere or political society is in deeper trouble.

Crises of a nonecological kind are strikingly apparent in the domain of electronic information systems, an area of technological change usually thought to be very progressive. Here the gap between the promise of these systems and the actual influence has produced three enduring paradoxes of the Information Age.

The Paradox of Intelligence

While information-processing machines are becoming more "intelligent" by leaps and bounds, much of the world's population appears to be moving in the opposite direction. Computers can now handle several billion calculations per second, microchip manufacturers pack ever-more computing capacity into ever-smaller spaces, and computing and telecommunications systems infuse our society to a degree unimaginable only a few years ago. On this front, horizons of development seem limitless. In stark contrast, there are growing indications that large segments of the population are deficient in mastering even the most basic intellectual skills. Hence, there has been a great outcry during the past decade that there is an emergency in our schools, worrisome indications of falling achievement test scores and a decline in basic verbal and mathematical skills of American students. A common response has been to attack the malady with a blitz of electronic information, spreading computers throughout the schools, in the hope that this would provide a remedy. But after a decade or so in which computer education has been applied, the signs of deterioration in both the schools and in the abilities of American school children each year appear unabated. In 1990 the national average of verbal scores on the Scholastic Aptitude Test (SAT) dropped two points. Reports of chaos in the schools—including violence and drug use—intensify with the publication of each new national study.

Of course, one can debate the origins and cures of this problem, but it's clear that as the Information Age matures, growing numbers in our population will approach the world in a state of increasing incompetence and bewilderment. Many institutions simply assume this and exploit the situation as an opportunity: in many fast-food restaurants you will not find numbered keypads on the cash registers, but pictures of hamburgers, french fries, and milkshakes. Social policy decisions, in which spending on education and job training becomes less important, mirror these hardware and software specifications.

A central theme in the time-honored ideology of progress, the belief that technological development and the enhancement of human abilities move forward together, is now effectively undermined by innumerable systems that successfully decouple these two ends through design programs that assume most working people are incompetent. A recurring pattern in modern technological and cultural transformation is that, as new technologies are invented, the kinds of people who will be using them are also invented. The creation of social roles and heavily promoted identities for the industrial factory worker, modern housewife, busy corporate manager, and suburban, car-driving, teenage consumer are among the more prominent American examples. As the century draws to a close, the invention of many new roles and identities presupposes a rapid dumbing-down of the world's populace.

The Paradox of Lifespace [193]

Most recent technologies have promised opportunities for greater leisure, creativity, and freedom. Without doubt, our social landscape is being transformed by fax machines, beepers, cellular phones, laptop computers, modems, e-mail, and numerous other devices and systems—all of which facilitate productivity. Any benefits are purchased at a definite cost. Corners of our lives formerly sheltered from direct technological intervention are now bombarded by the insistent call of incoming and outgoing messages. We will have to get used to public spaces—shops, restaurants, taverns, theaters, galleries—filled with people chattering on portable communications systems. Would you prefer the phoning or nonphoning section?

The automobile is undergoing similar transformations, fast becoming a vehicular multimedia center. How much more efficient it is to sail down the highway making deals, checking the stock market, or updating one's calendar! As one Los Angeles executive explained to a reporter from the *Wall Street Journal*, "I can't drive and enjoy the radio. I have to be on the phone." Consider the computerized systems of the "electronic cottage" and the evolution of homes into domestic work stations. Our society has begun to look like a vast electronic beehive in which information processing in search of economic gain overshadows other personal and social goods. Places and spaces in our lives formerly devoted to sociability, intimacy, solitude, friendship, love, and family are now being redefined as susceptible to productivity, transforming social norms and boundaries. Subjected to the pace of productivism, we come to think that if a message can move, it must move quickly. This places strong demands on individuals; communications technologies not only make it possible to reach them but obligates them to remain accessible.

The relentless introduction in the twentieth century of "time-saving" devices has encouraged us to think we would be liberated from toil, freed to pursue more creative work, peaceful reflection, more enjoyable sociability. These dreams have been inevitably frustrated. Our available time expands into a space of congestion—increasingly frenetic interactions encouraged by our machines. Although we "save time," we

have not been clever enough to bank it. At bottom is the bitter irony that the desire to do things faster and faster actually destroys the possibility that we might accomplish our tasks in a leisurely way. For, as we fill life's every niche with high-tech gadgetry, we gradually whittle away those quiet, restful places where genuine creativity and satisfaction are nurtured.

The Paradox of Electronic Democracy

Manifestos and ideologues of the Information Age have typically proclaimed that new information technologies will enhance democracy by better informing a self-governing citizenry and help us better connect to world events as well as to the knowledge necessary to act in socially and politically useful ways. For a small stratum of our society, those who hang out in transnational computer networks (likely to remain a small and privileged minority), this dream may seem close, but the information-processing machine that still matters most to the majority of people is television, although its younger progeny, virtual reality and interactive media, seem poised to leave the development labs for the mass market. This astonishing retreat into fantasy and sensationalism would not be so disturbing were it a temporary aberration or short-term fashion—if one can believe survey data, Americans watch an average of seven hours of television per day, roughly 665 billion waking hours a year in what seems a permanent pattern, an expanding cultural addiction. People do not seem to remember much of what they watch, but they need a certain daily dose of the TV drug in order to feel satisfied.

These video habits seem subject to a fascinating developmental logic wherein images and messages are broken down into ever-smaller pieces. On MTV, for example, where the vivid images on the screen change from eighty to one hundred times a minute or more, viewers have become acclimated to a frantic pulse and become impatient with ideas or arguments delivered in any less hurried way. Public discourse now assumes a profound fragmentation in the way people process information. American political leaders and their professional "handlers" are convinced any idea longer than a sentence is too difficult for the television-viewing public to endure. An increasingly obvious hollowing of public discourse has begun to undermine communication between office-holders and the general populace, producing a mood of embittered cynicism among many citizens. As American politics

[195]

has developed into a politics of video imagery, it has also rapidly devolved into a politics of disgust and endless complaint, linked to a widespread perception that politicians just do not know much or care much about ordinary people. Earlier in this century fascism and its propaganda masters fed on similar sentiments. For those who gleefully predicted public enlightenment and a renaissance of democracy in the Information Age, these are sobering developments.

Artifact/Ideas and Ventriloquism

The paradoxes outlined here suggest an enormous gap between commonplace expectations about the role information technologies ought to play in our lives and the role they have actually assumed. In the vicinity of those fabulous chips, circuits, tubes, and cables, one finds an astonishing deterioration of individual abilities, social spaces, and political practices. This isn't to say the new electronics have determined anything about the social consequences we observe. At the root, rather, are the power-holding institutions and processes that promote the domination of one or another technosocial form; in other words, combined patterns of material forms and social activities that comprise the "technologies" in question. Only after these patterns become widespread can one speak of their conditioning effects.

Beyond the prevailing progressivist ideology of technology lies the attempt to illuminate technology as a sphere of choices and conflicts. As one regards prospects for structuring new devices, techniques, and systems, one fruitful strategy is to notice how technologies embody ideas, what might be called artifact/ideas. As a person encounters a device or system, whether one in use or one on the drawing board, it is crucial that he or she ask what the form of this thing presupposes about the people who will use it. Having asked that question, one can move on to make explicit what artifact/idea or ideas the object embodies, that is, to give voice to the presuppositions in human-made things. In this form of artifactual ventriloquism, a McDonald's cash register might announce its underlying theme: "The system is intelligent; its users are not." Of course, those who try to speak the ideas embedded in material things are bound to disagree among each other, but if we are to overcome the silence that now surrounds the politics embedded within various technologies, it is crucial to take the first step, to speak prominent (but often covert) artifact/ideas right out loud.

Having begun such a dialogue with and about material things, we can go on to ask what technical devices and systems should presuppose about human beings. What forms and features should be present in the technologies our society adopts? A discussion of this kind allows us to explore the explicit and covert ideas that inevitably reside in objects, ideas about membership, gender, social class, access, control, order, community, freedom, and many more.

In this light, it is entirely appropriate that ideologies of technology should become a prominent topic for discussion among artists; for, art exists in the realm of artifact/ideas and helps give voice to material things. Clearly, any remedy to the crisis in political culture we are experiencing today would require a revitalization of the arts, the civic arts in the broadest sense. Beyond our century's threadbare slogans and technological revolutions stands the promise of reconstituting citizenship through playful experimentation with evolving social and technical forms. Will contemporary citizens, artists, and engineers respond to this challenge? Or will they follow the path usually held out to them, locating comfortable sinecures within the rapidly emerging transnational, technocorporate order, embracing and promoting ever more exquisite varieties of electronic addiction?

Notes

Margaret Morse

I wish to thank Jude Milhon (St. Jude of *Mondo 2000*), Mark Rennie of Smart Products, Sharon Grace, Jack Walsh, Steve Fagin, Danielle Escalera, and Alisa Steddom for their insights and suggestions, without implying that they approve of my conclusions. In the interest of cogency and space, this paper excludes remarks made at that forum drawn from a prior work on fast and fresh cuisine, "Telefood and Culinary Postmodernism." The latter will appear as a chapter in my *Television Reality: The Discursive Formats and Genres of Everyday Television*, forthcoming from Indiana University Press.

1. Epigraph on a nineteenth-century human growth hormone from chapter one of Lewis Carroll's *Alice in Wonderland*. The outcome of the pill is revealed in chapter two: "'Curiouser and curiouser!' cried Alice (she was so much surprised, that for the moment she quite forgot how to speak good English). 'Now I'm opening out like that largest telescope that ever was! Good-bye feet!'" Alice's early Wonderland adventures are largely culinary and, like those of smart drug users, especially concerned with "fit."
2. Allucquére Roseanne Stone has nominated this attitude "cyborg envy" in her article "Virtual Systems," in *Incorporations*, ed. Jonathan Crary and Sanford Kwinter (New York: Zone Books, 1991), pp. 609–621, which addresses the notion of "decoupling" agency or subjectivity from the physical body and calls attention to the spiritual overtones in virtual worlds. Yet, she claims, "The 'original' body is the authenticating source for the refigured person in cyberspace: no 'persons' exist whose presence is not warranted by a physical body back in 'normal' space. But death in either normal space or cyberspace is real, in the sense that if the 'person' in cyberspace dies, the body in normal space dies, and vice versa."

Manuel De Landa's *War in the Age of Intelligent Machines* (New York: Zone Books, 1990), a history of technology from the imagined standpoint of machines, implies not only a fundamental problem of incompatibility but ultimately of opposed interests of humans and machines. Of course, the discursive strategy of posing a subjectivized and empowered telos for machines is making humans aware of our own alienated, and thus unchecked, desires and actions.
3. Definition from *The American Heritage Dictionary*, cited in Gabriele Schwab, "Cyborgs: Postmodern Phantasms of Body and Mind," *Discourse* 9 (Spring–Summer, 1987), p. 80. Donna Haraway writes, "By the late twentieth century, our time, a mythic time, we are all chimeras, theorized and fabricated hybrids of machine and organism; in short, we are cyborgs. The cyborg is our ontology; it gives us our politics. The cyborg is a condensed image of both imagination and material reality, the two joined centers structuring any possibility of historical transformation." See Haraway, "A Cyborg Manifesto: Science, Technology and Socialist-Feminism in the Late Twentieth Century," in *Simians, Cyborgs and Women: The Reinvention of Nature* (New York: Routledge, 1991), p. 150. Haraway's "The Actors Are Cyborg, Nature Is Coyote, and the Geography Is Elsewhere: Postscript to 'Cyborgs at Large,'" in *Technoculture*, ed. Constance Penley and Andrew Ross (Minneapolis: University of Minnesota Press, 1991), pp. 21–26, insists on the metalepsis of the cyborg as "monster" or "boundary creature" who speaks from within the belly of the monster.

Of course, Haraway's is but one of many formulations of the cyborg metaphor—from *Robocop* and *The Terminator* versus *Terminator 2* to the totalizing subject of the war machine to Haraway's own bad-girl hybrid in a local war against dualisms. Her rhetorical strategy is to displace a female subjectivity from (a feminist essentialist vision of) nature *into* the machine, as opposed to De Landa's regrouping of the human/machine dichotomy and the subsumption of human subjectivity from the developmental logic of the machine.

Note that the "machine" metaphor has been applied to monastic life and other political and social organizations; such an application of the metaphor does not address the more literal problem of physical accommodation of the body and the electronic discussed here. However useful the cyborg as metaphor may be, it begs the question of fusion in the first place — or it is satisfied with making all tool users into cyborgs.

4. "Meat" is the human body as defined in *Mondo 2000*'s *A User's Guide to the New Edge* (New York: Harper Perennial, 1992), p. 170: "This expression communicates the frustration that people dealing with an infinitely expandable infosphere feel at the limitations imposed upon the wandering mind by the demands of the body."

5. See Elizabeth A. Grosz, "Lesbian Fetishism," *differences* 3, no. 2 (1991), pp. 39–54, for a lucid description of the difference between repudiation, negation, and disavowal as psychical defense mechanisms and an application of these categories strategically, rather than therapeutically, in cultural criticism. Grosz's article was brought to my attention by Patricia Mellencamp's *High Anxiety*.

6. The excerpt from Patterson's *Eating the "I": In Search of the Self* (San Anselmo, Calif.: Arete, 1992) continues: "I had always thought that one class of beings eats another; that all forms of life, gross and subtle, were engaged in a kind of perpetual eating or, as Gurdjieff called it, 'reciprocal maintenance.' The different classes (the vertebrates, invertebrates, man and angels) are separated by what they eat, the air they breathe and in what medium they live. It had never occurred to me that within classes of beings the strong psychically feed on the weak. The waitress was 'food' for her boss (p. 284)."

7. In identification, or mirroring and mimicry, one mistakes not-self (for instance, a mirror image) for the self. Thus, identification with a double or like demands that latter (ultimately to be overlooked) and a slight difference in scale between a subject and an object (a difference to be ignored).

8. Jean Laplanche and J. B. Pontalis, *The Language of Psycho-Analysis*, trans. D. Nicholson-Smith (New York: Norton, 1975), pp. 211–212. An oral typology of covering or engulfing/ being covered or engulfed also plays a role in the logic of some video games and computer displays and interaction with, for instance, the Pacman or Macintosh windows and interface.

9. Ibid., p. 287. The passage specifically refers to "the love-relationship to the mother." In some psychoanalytic thinking, orality is not restricted to an oral or primitive stage. In "Rethinking Cannibalism," Richard M. Gottlieb (New York University, New York, 18 August 1992) notes the pervasiveness of cannibalistic fantasy in contemporary culture. His remarks especially emphasize the theme of the body in pieces (much like food that is cut up, broken, and torn) as well as the theme of the resurrected or intact body. Gottlieb also suggests that digestion, decay, and decomposition are part of this body as food continuum, themes comparable to the terms of *waste* and *abjection* employed here.

10. J. G. Ballard, "Project for a Glossary of the Twentieth Century," in *Incorporations*, ed. Crary and Kwinter, p. 277.

11. For contrasting visual images of nonfood, see "Future Food," a glamour shot of gels and pills, steel worm/conduit, and rather menacing looking brussels sprouts (year 2010), styled by Erez and photographed by Joshua Ets-Hokin, in *Mondo 2000* 3 (Winter 1991), p. 102. A contrasting matte and ascetic simulation of an ad for a soup of supplements captioned, "Do you need a soup of supplements?…especially if you can't stand broccoli and brussels sprouts," appears in Anastasia Toufexis, "Health: The New Scoop on Vitamins," *Time* 6 April 1991, pp. 55–59.

12. Metaphor courtesy of St. Jude, *Mondo 2000*.

13. Faith Popcorn, *The Popcorn Report* (New York: Bantam, 1991), p. 66f. Durk Pearson and Sandy Shaw of Designer Foods™, however, emphasize that "All ingredients [in their SMART Products] are recognized and utilized by the human body as a food," *Intelli-Scope: The Newsletter of Designer Foods™ Network* 2 (February 1993), p. 4. Despite its name, the newsletter also denies that "Smart Drinks" will improve IQ, even as it affirms that they will increase well-being, energy, and clarity, especially for someone who is "feeling constantly tired, drained or stressed out."

14. Durk Pearson and Sandy Shaw, "Durk and Sandy Explain It All for You," *Mondo 2000* 3 (Winter 1991), p. 32.

15. Karl Abraham set the oral-sadistic stage concurrent with teething and the fantasy of being eaten or destroyed by the mother, whereas Melanie Klein placed infantile sadism throughout the oral stage. "The libidinal desire to suck is accompanied by the destructive aim of sucking out, scooping out, emptying, exhausting" (quoted in Laplanche and Pontalis, *The Language of Psycho-Analysis*, p. 289).

16. Hans Moravec, "The Universal Robot," in *Out of Control: Ars Electronica 1991*, ed. Gottfried Hattinger and Peter Weibel (Linz: Landesverlag, 1991), p. 25.

17. Quoted in Jeremy Rifkin, *Biosphere Politics: A New Consciousness for a New Century* (New York: Crown, 1991), p. 246; from Grant Fjermedal, *The Tomorrow Makers* (New York: MacMillan, 1986), p. 8. Moravec and Sussman appear to entertain the questionable idea that there will still be an "I" and a "mind" capable of "thought" in this future union of brain and hardware.

18. Cited in the publicity for the exhibition "Ars Electronica '93: Genetic Art. Artificial Life" in Linz, Austria.

19. Compare this to Andreas Huyssen's reasoning in *After the Great Divide: Modernism, Mass Culture, Postmodernism* (Bloomington: Indiana University Press, 1986), or Freud's explanation of the death drive in *Beyond the Pleasure Principle*, in *The Standard Edition of the Complete Psychological Works of Sigmund Freud*, vol. 18 ed. James Strachey (London: Hogarth, 1953–1974), pp. 3–64.

20. For a perceptive application of Klein's psychoanalytic theories to video games, see Gillian Skirrow, "Hellivision: An Analysis of Video Games," in *High Theory, Low Culture*, ed. Colin MacCabe (Manchester: Manchester University Press, 1986).

21. I am not unaware of the metaphor of oral sex as "eating," but here eating is not a metaphor, and sexuality reappears in terms of sadomasochistic affect—as pain and ecstasy. Flanagan "explores the effects of chronic illness on sexual identity, in particular, sadomasochism" in his installation *Visiting Hours* at the Santa Monica Museum of Art, LAX Festival 1992. *Bob Flanagan: Super Masochist* (San Francisco: Re/Search, 1994) appeared after the completion of this essay.

22. Tzvetan Todorov, *The Conquest of America*, trans. Richard Howard (New York: Harper, 1985), p. 158.

23. SmartSkin™ is a new product introduced in the *Intelli-Scope* (cited above). For a psychoanalytic elaboration of the concept of the skin ego (especially the second skin) with case histories, see Didier Anzieu's *The Skin Ego: A Psychoanalytic Approach to the Self*, trans. Chris Turner (New Haven: Yale University Press, 1989), in particular the chapters on "The Second Muscular Skin," "The Envelope of Suffering," and "The Film of the Dream." Anzieu is also enlightening on the psychological significance of breaking the skin in sadomasochistic rituals. While *The Skin Ego* does not deal with electronic skin, it does discuss film. Two striking references to skin in film occur in the segment "Hoichi, the Earless One" in Kwaidan, in which holy words written on the protagonist's skin protect his body against being eaten by ghosts; a more recent reference occurs in *The Silence of the Lambs*, both the novel and the film. But, more importantly, Anzieu's model leads to the experience of projected film itself as enveloping sound and patterns of light and darkness.

24. I imagine such a situation in real space, once projected images are freed from boxes or frames, in my essay "The End of the Television Receiver," in *From Receiver to Remote Control: The TV Set*, ed. Reese Williams and Matthew Geller (New York: New Museum of Contemporary Art, 1990), pp. 139–141.

25. Todorov, *The Conquest of America*, p. 158.

26. Jean-Paul Sartre, "Intentionality," trans. Martin Joughin, reprinted in *Incorporations*, ed. Crary and Kwinter, p. 387.

27. As in, "His eyes devoured her" (ibid., p. 390). Julia Kristeva's critique of the philosophy of "meaning as the act of the *transcendental ego*, cut off from its body, its unconscious, and also its history" ("The System and the Speaking Subject," *Times Literary Supplement* 12 October 1973, pp. 1249–1250) refers to Klein and could as easily refer to virtual reality.

28. Deleuze and Guattari describe the development of "face" as a "white wall, black hole system." See Gilles Deleuze and Félix

Guattari, *A Thousand Plateaus: Capitalism and Schizophrenia*, trans. Brian Massumi (Minneapolis: University of Minnesota Press, 1987), pp. 167–191. This system is implicitly racially determined, as Meaghan Morris has suggested in her essay, "Great Moments in Social Climbing: King Kong and the Human Fly," in *Sexuality and Space*, ed. Beatriz Colomina (New York: Princeton Architectural Press), pp. 1–51.

29. Yet as malnutrition, hunger, and deficits have grown, disentitlement has become an increasingly respectable political option. Many different kinds of theorists, from anthropologists like Mary Douglas to political scientists and activists suggest that hunger is most often a result of lack of legal entitlement and social inequities rather than of scarcity.

30. PBS, "All Things Considered," 10 December 1992.

31. Dr. Robert Klesges put on a tape of the television dramedy "The Wonder Years" and measured the metabolic rate of thirty-two girls ranging from seven to twelve years old, half of them normal weight, half of them obese. All showed a drop in metabolic rate, but that of the obese children was especially striking, suggesting why heavy use of television and obesity go together. The study was cited by Jane E. Brody, in "Literally Entranced by Television, Children Metabolize More Slowly," *New York Times*, 1 April 1992.

32. Michael Sorkin, "Scenes from the Electronic City," *ID* (May–June 1992), pp. 75, 77.

33. Molly O'Neill, "Watch for a Few Good Meals," *New York Times Magazine* 3 June 1993, p. 29.

34. Sharon Zukin links nouvelle cuisine and gentrification as "restructurings of socio-spatial power" in "Gentrification, Cuisine, and the Critical Infrastructure: Power and Centrality Downtown," in *Landscapes of Power: From Detroit to Disney World* (Berkeley: University of California Press, 1991), pp. 179–215.

35. One San Francisco ex-restaurateur and co-owner of the mail-order firm Smart Products ™ says "I don't eat breakfast anymore. I don't eat lunch; I drink smart drinks" that take one minute to stir and one

to consume. He would welcome being able to abandon "real" food from depleted and (we might add, even toxic) soils entirely. See Mark Rennie, "Smart Drugs' True Believers: Highly Developed Thoughts on These Additives for the Psyche," *San Francisco Chronicle* 4 March 1992, and telephone interview with the author, 2 April 1992.

36. I am basing my discussion here on the useful distinctions offered by Elizabeth Grosz in "Lesbian Fetishism." (cited above).

37. "The psychotic's hallucination is not the return of the repressed, that is, the return of a signifier, but the return of the Real that has never been signified—a foreclosed or scotomized perception, something falling on the subject's psychic blind spot. The subject's perception is not projected outward onto the external world. Rather, what is internally obliterated reappears for the subject as if it emanates from the Real, in hallucinatory rather than projective form. It confronts the subject from an independent, outside position." Grosz, "Lesbian Fetishism," p. 45f.

38. Stone, "Virtual Systems," p. 609n2.

39. William Gibson, *Neuromancer* (New York: Ace, 1984), p. 67. See also: Case "curled in his capsule in some coffin hotel, his hands clawed into the bedslab, temperfoam bunched between his fingers, trying to reach the console that wasn't there" (p. 5). And: "Case sat in the loft with the dermatrodes strapped across his forehead....Cowboys didn't get into simstim, he thought, because it was basically a meat toy. He knew that the trodes he used and the little plastic tiara dangling from a simstim deck were basically the same, and that the cyberspace matrix was actually a drastic simplification of the human sensorium, at least in terms of presentation, but simstim itself struck him as a gratuitous multiplication of flesh input" (p. 137f).

40. Ibid., p. 67.

41. Ibid., pp. 137f, 138.

42. Grosz, "Lesbian Fetishism," p. 45.

43. Calvin Sims, "Junk Food That's Lean and Healthy?" *New York Times* 27 January 1988.

44. Benjamin Woolley, *Virtual Worlds: A Journey in Hype and Hyperreality* (Cambridge, Mass.: Blackwell, 1992).

[201]

45. Ibid., p. 254.

46. Woolley's conclusion: "One of the aims of this book has been to show that reality is still there, though not in the material realm of the physical universe where the modern era assumed it to be. In my attempt to distinguish between simulation and imitation, the virtual and the artificial, I have tried to provide a glimpse of where that reality may be, in the formal, abstract domain revealed by mathematics and computation. This is not to say that any mathematics can discover it. Rather, the computer has, through its simulative powers, provided what I regard as reassuring evidence that it is still there." (Ibid., p. 254.) In a contrasting view, Paul Virilio castigates the very will to mathematical power as the occasion for voluntarily blinding the horizon of sight and hearing, in *La machine de vision*, German trans. *Die Sehmaschine* (Berlin: Merve, 1989), p. 171.

47. Naimark's "Nutrition" was made with students at the San Francisco Art Institute, 1990. A prior piece, *Eat* (1989), explores the absurdity of a virtual restaurant, with video projection meals like "Jackson Pollock" and an "eat" button that can lead to surprising effects—like the Heimlich maneuver.

48. Disavowal allows two contradictory forms of defense to coexist without influencing each other: "It does not rely on the unconscious, but pre-dates it. Like repudiation, it involves a split in the ego, but does not involve a failure in representation. The child's acceptance and refusal of reality… generates the representational impulse to produce profuse significatory contexts and fantasy scenarios" (Grosz, "Lesbian Fetishism," p. 46).

49. It is true that the virtual reality interfaces of glove, helmet, or suit map the physical body onto the virtual body—but not in a one-to-one correlation. The immersive illusion depends on kinesthetic sensations in the real body as it moves, but the machine tracking body coordinates allows for locomotion in a very circumscribed area in real space. Descriptions of the body in virtual reality are contradictory—probably because there are at minimum two bodies, one virtual and one actual. The virtual body can be delegated to one or more figures in the artificial world or remain subjective. The actual body can reportedly feel "disembodied." Scott Fisher imagines a man in the year 2001 composing a model for a new pharmaceutical in virtual space: he is crawling on all fours, as if "miming a wrestling match with himself, or recapitulating his infancy." (Scott S. Fisher, "Virtual Interface Environments," in *The Art of Human-Computer Interface Design*, ed. Brenda Laurel [Reading, Mass: Addison-Wesley, 1990], p. 423f). Interfaces like the helmet and data suit that occlude impressions of the physical world are, according to Myron Krueger, a problem; in remarks made at the Cyberthon, he offered a simple solution—transparent lenses of virtual information beyond which the physical world may be viewed.

50. In terms of description of effects, Ward Dean, M.D., and John Morgenthaler, for instance, report that piracetam promotes the flow of information between the right and left hemispheres and "might increase the number of cholinergic receptors in the brain. Older mice were given piracetam for two weeks and then the density of the muscarinic cholinergic receptors in their frontal cortexes was measured. The researchers found that these older mice had thirty to forty percent higher density of these receptors than before. Piracetam, unlike many other drugs, appears to have a regenerative effect on the nervous system." (Ward Dean and John Morgenthaler, *Smart Drugs and Nutrients: How to Improve Your Memory and Increase Your Intelligence Using the Latest Discoveries in Neuroscience* [Santa Cruz, Calif.: B&J, 1990], p. 44). Deprenyl is described by Morgenthaler, "Smart Drugs Update" *Mondo 2000* 5 (Summer 1992), p. 36.

Ross Pelton's *Mind Food and Smart Pills* (New York: Doubleday, 1989), and Pearson and Shaw's *Life Extension: A Practical Scientific Approach* (New York: Warner, 1983) and *The Life Extension Companion* (New York: Warner, 1984) offer more detailed explanations of these drugs and lengthier citations of references than does *Smart Drugs*. St. Jude's tongue-in-cheek article "Are You as Smart as Your Drugs? A Paranoid Rant by St. Jude," *Mondo 2000* 5, about smart-drug addiction ("Let's just face it—nobody with any chemically evolved intelligence is going to be exactly chuffed about, sinking back into the primordial ooze, brain-wise"), describes the efficacy of "800 mg of pirac-etam" in remembering "an obscure Japanese technical term — *kyogen*! that's the word — never noticing how all-out *unnatural* it is to recall stuff like *kyogen*" (p. 38). In a telephone interview Jude, who is both knowledgeable and skeptical about main-stream medicine, revealed her past as a physi-cian's assistant. Her pseudonym is an indica-tion of the knowledge of the hopeless, yet nevertheless...

51. Pelton, *Mind Food*, p. 318.

52. Roland Barthes, *Empire of Signs*, trans. Richard Howard (New York: Hill and Wang, 1982).

53. Toufexis, "Health," p. 56. In the side bar to the *Time* piece we read, "Can You Survive by Pills Alone?"

54. Pearson and Shaw, "Durk and Sandy Explain It All for You," p. 32. Like Alice, the discourse of smart drugs and drinks is all about "too much" and "too little." Terence McKenna, in *Food of the Gods: The Search for the Original Tree of Knowledge, a Radical History of Plants, Drugs, and Human Evolution* (New York: Bantam, 1992), also posits a dis-tinction between bad drugs (which include television) and good drugs (largely hallucino-genics) as one of "fit" fostered by coevolu-tion of plants with shamanistic practices from archaic times.

55. Pearson and Shaw, *Life Extension*, p. 168.

56. McKenna, *Food of the Gods*, p. 24.

57. Ibid., p. 274.

58. Dean and Morgenthaler, p. 206.

59. Testimonials from ibid., pp. 179–184, and Morgenthaler, "Smart Drugs Update."

60. See especially Pearson and Shaw, *Life Extension*, but this theme is universal.

61. St. Jude, "Are You as Smart as Your Drugs?," p. 38.

62. Julia Kristeva, *Powers of Horror: An Essay on Abjection*, trans. Leon S. Roudiez (New York: Columbia University Press, 1982), p. 3.

63. Ibid., p. 4.

64. *Abject Art: Repulsion and Desire in American Art*, exhibition catalogue (New York: Whitney Museum of American Art, 1993), was published after the completion of this essay. It argues a broader case for abjec-tion in art than is proposed here.

65. Christine Tamblyn treats the abject in the work of feminist performance artists like Karen Finley in "The River of Swill: Feminist Art, Sexual Codes and Censorship," *Afterimage* 18, no. 3 (October 1990), pp. 10–13.

66. Helms sought to amend Public Law 101–121 of 1989, banning the NEA from funding obscene art to include the depiction or description of sexual or excretory activi-ties or organs "in a patently offensive way," a prohibition that one senator, Tim Wirth (D–Col), noted medically includes the human skin. According to Helms, much of the art sanctioned by the NEA "turns the stomach of any normal person" (*New York Times* 20 September 1991). In the words of Eric Planin of the *Washington Post* (20 September 1991): "Seemingly unimpressed by the NEA's get-tough policy in denying grants to artists who, for instance, smear chocolate on their nude bodies or urinate on stage, the Senate voted 68 to 28 to adopt the amendment." This Senate stand was later abandoned in a "Corn for Porn" swap, that is, the maintenance of cheap grazing rights on public lands. Subsequent events — the charges of "blasphemy" by Pat Buchanan's presidential primary campaign ad attacking Marlon Riggs's *Tongues Untied* (a lyrical documentary on black, homosexual identity politics), as well as the reportedly forced resignation of John Frohnmeyer, NEA chair, resulting from the pressure of the Buchanan presidential campaign on President George Bush — support the role of "excretory art" as site of struggle over cultural change.

67. Barry chose two models who bore an uncanny resemblance to each other. The photographic results were digitally com-bined, as well as manipulated in other ways: since the fit between a human head, the severity of a cube, and the aspect ratio (length in relation to width) of a video screen is not ideal, judicious computerized stretching of the digital image was required, especially around the hairline, to make the head fit the cube to the corners. The result is the restoration of at least part of the body to art as minimalist sculpture.

68. Charles Hagen compares *Imagination Dead Imagine*, "With its blend of the sensual and the repulsive," explicitly with "Andres Serrano's photography combining religious symbols and bodily fluids" in his 25 October 1991 review in the *New York Times*. Barry cites other literary inspirations—for instance, J. G. Ballard's "The Impossible Room," in *The Atrocity Exhibition* (San Francisco: Re/Search, 1990), includes the following description: "A perfect cube, its walls and ceilings were formed by what seemed to be a series of cinema screens. Projected on to them in close-up was the face of Nurse Nagamatzu, her mouth, three feet across" (p. 33).

69. Words that embody or enact what they mean are in fact calligrams—except that "calli" means beautiful. Michel Foucault wrote on the phenomenon in "Ceci n'est pas une pipe," *October* 1 (1976), pp. 9–10.

IV

Technology, Art, and Cultural Transformation

Artists, Engineers, and Collaboration

Billy Klüver

One of the most persistent ideas in twentieth-century art is that of absorbing new technology into art: the Futurists' blind devotion to technology, the Russian Constructivists' attempts to merge art and life into new imaginative forms, the more rigorous design approaches at the Bauhaus, continued by Gyorgy Kepes at MIT, and the work of individual artists such as Marcel Duchamp and John Cage. This involvement with technology has represented artists' positive desire to be engaged in the physical and social environment around them.

In the early 1960s, when technology began to develop rapidly, many artists wanted to work with forms of new technologies, but often found themselves shut out, with little or no access to technical and industrial communities. When, in 1960, I began to collaborate with artists on their projects, I was working as a scientist in the Communication Sciences Division at Bell Telephone Laboratories and had virtually unlimited access to technical people and resources, and most importantly, I had the tacit support of executive director of the division, John R. Pierce. I will discuss the evolution of these one-on-one collaborations between artists and engineers, and their development into the foundation Experiments in Art and Technology (E.A.T.).

Jean Tinguely came to New York City in early 1960. On seeing the city for the first time, he decided to build a large machine that would violently destroy itself in front of an audience in a theater, throwing off parts in all directions. A protective netting would save the audience. When the Museum of Modern Art invited Jean to build his machine in the garden of the museum, he asked me for help. I took him to the New Jersey dumps, which in those days were not covered with dirt. He found bicycle wheels, parts of old appliances, tubs, and other junk, which we hauled to the museum and threw over the fence into the garden.

[207]

Jean Tinguely, *Homage to New York* (1960), photograph: David Gahr.

Enlisting the help of Harold Hodges at Bell Laboratories, we built a timer that controlled eight electrical circuits that closed successively as the machine progressed toward its ultimate fate. Motors started; smoke, generated by mixing titanium tetrachloride and ammonia, bellowed out of a bassinet; a piano began to play and was later set on fire; smaller machines shot out from the sculpture and ran into the audience. In order to make the main structure collapse, Harold had devised a scheme of using supporting sections of Wood's metal, which would melt from the heat of overheated resistors. The whole thing was over in twenty-seven minutes. The audience applauded, and then descended on the wreckage for souvenirs. Jean called the event *Homage to New York*. During those three or four weeks of the construction of the machine, I learned how to listen to the artist, and to give him as many technical choices as I could—as quickly as possible. And as Jean has said repeatedly since, it couldn't have happened without our collaboration.

Shortly thereafter, Robert Rauschenberg asked me to collaborate on what he described as an interactive environment, where the temperature, sound, smell, and lights would change as the audience moved through it. After many discussions, the idea boiled down to a sound environment where the sounds came from five AM radios. From a central control unit, the audience could vary the volume and the rate at which the AM band of each radio was being scanned. But Bob wanted no wires between the control unit and the radios. Considering the electronics available in the early sixties, this turned out to be a difficult technical problem. We designed a system in which all the AM radios were located in the control unit and the sound was retransmitted on FM to receivers and speakers. We had a lot of trouble with interference between the AM receivers and with noise from the small motors that drove the scanners. When we solved these problems, Bob put together the five sculptures that make up *Oracle* from objects he found in the streets; and the control panel, receivers, and speakers were installed in them. *Oracle* is now at the Centre Georges Pompidou in Paris. The technology has now been updated for the fourth time, using electronic scanning and infrared transmission between the pieces. After thirty years the technology has finally caught up with the artist, and *Oracle* is performing as it was originally conceived.

Robert Rauschenberg, *Oracle* (1965). Copyright © Robert Rauschenberg/VAGA, New York, photograph: Rudy Burckhardt.

Jasper Johns, *Zone* (1962), 60 x 36 inches.
Collection Kunsthaus Zurich. Copyright ©
1994 Jasper Johns/VAGA, New York.

Jasper Johns, *Field Painting* (1964),
72 x 36 3/4 inches. Collection of
the artist. Copyright © 1994 Jasper
Johns/VAGA, New York.

Jasper Johns asked if he could make a painting with a neon letter in
it. What was new was that Johns wanted no cords to the painting. We
needed a battery-powered high-voltage supply, but to stack up batteries
attached to seven hundred volts would have been messy, dangerous, and
impractical. So we started out with twelve volts of rechargeable batteries. A multivibrator circuit converted the DC voltage from the batteries
into AC. Transformed into seven hundred volts and then rectified, it
powered the neon letter. All the technical equipment was mounted
behind the painting. We were able to provide enough energy for the
blue "A" which sticks out horizontally at the top of the painting in
Zone (1962) and the red neon "R" in *Field Painting* (1964).

One day in the summer of 1964 in his Forty-Seventh Street studio, Andy Warhol asked me if we could make him a floating light bulb. My colleagues at Bell Laboratories and I made some calculations and discovered that it was not possible with existing battery technology. While working on the idea, another colleague found a material called Scotchpak, which was relatively impermeable to helium and could be heat-sealed. The United States Army used it to vacuum-pack sandwiches. Andy wanted to use the material to make clouds. While we were experimenting with how to heat-seal curves, Andy took the material, folded it over, and made his *Silver Clouds*. When they were shown at the Leo Castelli Gallery in April 1966, the heat gradient between the floor and the ceiling created a slight pressure differential, and with paper clips as ballast, we balanced them so that they would float halfway between the ceiling and the floor.

By 1965 I had taken dozens of artists through Bell Laboratories and many of my colleagues had worked with artists, but I began to feel a larger effort was necessary to increase the awareness of the technical

Andy Warhol, *Silver Clouds* (1966). Courtesy Leo Castelli Gallery/
The Andy Warhol Foundation for the Visual Arts, photograph: Rudy Burckhardt.

community and make it more accessible to artists. This became possible when a group of artists, many of whom had performed together at Judson Church, expressed a desire to stage large-scale performances in collaboration with scientists and engineers. Out of this came "9 Evenings: Theater and Engineering," a series of performances at the 69th Regiment Armory in New York City in October 1966 by ten artists: John Cage, Lucinda Childs, Oyvind Fahlström, Alex Hay, Deborah Hay, Steve Paxton, Yvonne Rainer, Robert Rauschenberg, David Tudor, and Robert Whitman. Each of the ten artists worked closely with one or more engineers, primarily from Bell Telephone Laboratories.

The first meeting of artists and engineers took place in early 1966 in Rauschenberg's studio. During the summer of 1966, more than thirty engineers were hard at work, with at least one engineer assigned to each artist, depending on the artist's project and engineer's specialty. For example, Bill Kaminski designed and built for Alex Hay low-noise differential amplifiers with 80db gain and FM transmitters that could pick up and transmit body sounds, muscle activity, eye movements, and brain waves from electrodes attached to his body. Peter Hirsch developed a

Lucinda Childs, "Vehicle,"
9 Evenings (October 13, 1966).
Photograph: Peter Moore.

Doppler sonar for Lucinda Childs. Three red buckets swung inside a simple scaffolding, on the periphery of which were mounted three seventy kHz ultra—high-frequency sound transmitters generating inaudible sound beams, which were reflected from the moving buckets. Through the Doppler effect, the reflected sound beam had a frequency slightly higher or lower than seventy kHz and the beat frequency between the return signal and seventy kHz, which was proportional to the speed of the buckets, was amplified and fed through the speakers in the armory. The resulting sound was like wind blowing through a forest.

Other engineers worked on equipment and systems that would be used by more than one artist; in particular, a local-area FM transmitting system used to control lights, sound, and movement of objects at a distance. Fred Waldhauer designed a proportional control system for moving sound around the speakers mounted in the armory and for varying the level of the sound in each speaker, which was used by John Cage, Deborah Hay, and David Tudor.

Robert Rauschenberg's work *Open Score* combined the FM transmitting system with elements unique to his piece. In the first part, Frank Stella and Mimi Kanarek played tennis. Each time they hit the ball, a small specially designed radio transmitter embedded in the racquet handle transmitted the vibration of the racquet strings to the speakers around the armory, and a loud bong was heard. For each bong, a light went out, and the game ended when the armory was in complete darkness. With the use of infrared light and infrared-sensitive television cameras, the images of the crowd as they moved in the space were projected on three large screens suspended in front of the audience. The audience could feel that the people were there but could not see them except on the screens. The infrared camera tubes came from Japan, since they were classified as secret by the military in this country.

Most of the equipment used in "9 Evenings" did not exist off-the-shelf in 1966 and was built especially for the artists by the engineers. All together, the engineers contributed four man-years of engineering to the performances. "9 Evenings" ran from October 14 to 23, 1966, and more than ten thousand people attended over the course of the performances.

"9 Evenings" raised enormous interest among New York artists in using new technology. Robert Whitman, Fred Waldhauer, Robert Rauschenberg, and I decided to form E.A.T., a service organization for artists, engineers, and scientists. Three hundred artists showed up at our first meeting in November 1966, and eighty made immediate requests for technical help. We began to actively recruit engineer members, published a newsletter, held open houses where artists and engineers could meet informally, and organized lecture-demonstrations by scientists for artists on topics ranging from lasers to computer graphics to paper to color theory. Within three years we had recruited more than two thousand engineers from all over the country and established a technical services matching system to put artists directly in touch with engineers. We made a conscious effort to help every artist who approached us with a request.

In late 1968 Pepsi-Cola approached E.A.T. about designing and programming a pavilion for Expo '70 in Osaka, Japan. The original four artists who began the collaborative design of the pavilion were Robert Breer, Robert Whitman, Frosty Myers, and David Tudor. As the design of the pavilion developed, engineers and artists were added to the project and given responsibility to develop specific elements. Finally, sixty-three engineers, artists, and scientists in the United States and Japan contributed to the design of the pavilion.

As the exterior and interior elements of the pavilion developed, so did the guiding notion of the pavilion. It became an ever-changing place where each visitor would be encouraged to explore and create an individual experience. The pavilion was designed as a performance space as well, continuously programmed by invited artists throughout the six-month duration of Expo '70.

[214] The visitor entered the pavilion through a tunnel and descended into a dark clam-shaped room, lit only by moving patterns of laser light from a sound-activated laser display system developed by Lowell Cross and David Tudor. The path continued upstairs into the main space of the pavilion, a ninety-foot diameter, 210-degree spherical mirror made of aluminized Mylar. The floor and the people moving on it were all reflected upside down as "real" images in the mirror. (A "virtual" image

Pavilion for Expo '70, Osaka, Japan (1970). Photograph: Harry Shunk.

is one you see "behind" a flat mirror; a "real" image appears in front of the mirror, roughly the same distance from the center of the sphere as you are on the other side of the center.) A "real" image produced in a spherical mirror resembles a hologram. Because of the size of our mirror, however, a spectator looking at the real image of someone in the mirror could walk around the image and see it from all sides. The space in the mirror was gentle and poetic, rich and always changing. It was visually complex and we discovered new and complicated optical effects every day. Once visitors could see themselves or their friends as three-dimensional real images in the mirror space, the reaction was incredible and created much more excitement than we ever could have expected.

David Tudor conceived of the interior of the mirror dome as a sound environment and designed the sound system as an "instrument" that could be programmed or played by visiting artists. Recognizing the unique properties of the spherical mirror, thirty-seven speakers were arranged in a rhombic grid on the surface of the dome behind the mirror. Sound could be moved from speaker to speaker at varying speeds linearly across the dome and in circles around the dome. It could also be shifted abruptly from any one speaker to any other speaker, creating point sources of sound.

Beneath all this, the floor was divided into ten sectors, each made of a different material, in which were embedded wire loops serving as antennae that transmitted a highly localized sound signal. Using handsets, visitors could hear sounds specific to each different floor material: on the tile floor, horses, hooves and shattering glass; on the Astroturf, ducks, frogs, cicadas, roaring lions, and so on.

Outside the pavilion, the dome-shaped roof was covered by a water-vapor cloud sculpture by the Japanese artist Fujiko Nakaya. The cloud was produced when water under high pressure was pushed through 2,520 jet-spray nozzles and broken up into water drops small enough to remain suspended in air. On the plaza in front of the pavilion seven of Robert Breer's *Floats*—six-foot high, dome-shaped sculptures—moved around at less than two feet per minute, emitting sound. At night Frosty Myers's *Light Frame* sculpture traced a well-defined tilted square of white light around the pavilion. Four three-legged black poles of different heights were set in a square at each corner of the pavilion plaza. Two high-intensity xenon lights were placed atop each pole. Each light was directed toward the light of the neighboring tower, and specially designed parabolic reflectors kept the light beams narrow, which defined the sides of the square of light.

The number of technical breakthroughs in the pavilion was quite astonishing; almost every system we designed was new and untried. But even more significant, the artists and engineers had created a living, responsive environment that was different for each visitor. Three million people visited the pavilion during the summer of 1970.

In the 1970s we became more interested in interdisciplinary collaborative projects that involved artists in other areas of society. The first grew out of a request from Vikram Sarabhai, head of the Indian Atomic Energy Commission, to develop procedures for producing instructional material to be broadcast from the AST-F satellite to hundreds of Indian villages. We put together a team that included engineers, artists, psychologists, and education specialists, and chose to work on instructional programs for women who owned water buffaloes in a dairy cooperative near Baroda in Gujurat state. The challenge was to preserve the local cultural component and overcome the built-in cultural aesthetics associated with instructional programming inherited from the West. We proposed that visual material be generated by the villagers themselves on such subjects as artificial insemination, proper nutrition for the buffalo, treatment of the diseases, and so on, using half-inch videotape; these tapes would then be bicycled to another village to be shown and evaluated. On the basis of this recorded material, the final programs on professional broadcast tapes would be made. Our proposal was in fact adopted for the SITE satellite educational television project and was carried over to other areas of instructional television in India.

"Children and Communication" was designed and run in collaboration with education specialists from New York University. Children from different neighborhoods in New York City became acquainted with each other through the use of various types of communication equipment, never having to leave their own neighborhoods. One center was set up on Sixteenth Street and one on Sixty-Eighth Street with open lines for telephones, telex machines, facsimile machines, and telewriters. Robert Whitman designed the physical environment for each center. Groups of children at each location freely used the equipment to communicate with each other. The project generated hundreds of drawings which depicted how the children saw the experience, and the hard copy from the telex and telewriter machines reflected the ingenuity and enthusiasm of the children in making contact with each other.

Another communication project, "Utopia Q and A," was part of Pontus Hulten's exhibition at Moderna Museet in Stockholm commemorating the hundred year anniversary of the Paris Commune of 1871. Using telex machines in Tokyo, Ahmedabad (north of Bombay), Stockholm, and New York, the public was able to send technical or opinion questions about ten years in the future—1981—to the other three countries. The answers—from experts or from the general public—were telexed back to the questioner. Hundreds of questions and answers were exchanged over the month-long operation of the project. The general tone of the Japanese questions and answers were optimistic; the American, more pessimistic; the Swedish, critical; and the Indian, theoretical.

E.A.T.'s contribution to the social dialogue of the 1960s and '70s was the idea of one-to-one collaborations between artists and engineers. E.A.T. opened up exciting possibilities for the artists' work by finding engineers willing to work with them in the artists' own environment. Together the artist and the engineer went one step beyond what either of them could have done separately. But perhaps more importantly, the artist-engineer collaboration was the training ground for larger-scale involvement in social issues for both the artist and the engineer. In the nonart projects that E.A.T. undertook, at least one artist was always part of the interdisciplinary team, and we put a high value on the expertise the artist brought to the project.

The "expertise" that artists bring to the collaboration comes directly from their experience in making art. The artist deals with materials and physical situations in a straightforward manner without the limits of generally accepted functions of an object or situation, and without assigning a value hierarchy to any material. The audacity of Picasso's collages in his time, Meret Oppenheim's surrealist objects, and Rauschenberg's combines and cardboard pieces all illustrate this quality. The artist makes the most efficient use of materials, and achieves the maximum effect with minimum means. Surplus of material leads to decorative work. The artist is sensitive to scale and how it affects the human being. From cave drawings to Persian miniatures, cathedral frescoes, or Christo's *Running Fence*, scale has been a consistent concern of the artist. The artist is sensitive to generally unexpressed aesthetic assumptions, which are based on subjective preference masquerading as "objective," practical, economic, or social factors. The artist assumes total responsibility for the artwork. The artist knows that a work is the result of personal choices; this sense of commitment and responsibility gives the artist and the work a unique quality.

The engineer, of course, brought to these collaborations technical expertise and an interest in problem solving. While the technology needed by the artists might often be "trivial" from the engineer's point of view, its application in a new environment for a new use provided difficulty and challenge. In Rauschenberg's *Oracle*, we had to build a multichannel FM broadcasting system in a single room!

Those of us in the technical community in the early sixties who were worried about the direction of technological change believed that artists' ideas, approaches, and concerns could influence the way engineers approach technological or day-to-day social problems. Our collaborations, we hoped, could lead technological development in directions more beneficial to the needs, desires, and pleasures of the individual.

An interesting comment on my experience in working with artists came from Nam June Paik, when he told me recently, "Billy, I am working with off-the-shelf technology, you always worked to invent one-of-a-kind technology." Paik, of course, was understating his extraordinary visual sense in manipulating his material, but he hit the nail on the head about the driving force in the interaction between artists and engineers: what will emerge is something that neither the artist nor the engineer had thought of before. Thus, the artist-engineer collaboration remains a viable model for how we can actively confront and shape new technology.

Stories from the Nerve Bible

Laurie Anderson

EAST: O Little Town of Bethlehem: rock-throwing capital
of the world.
WEST: those who came before me.
UP: the true meaning of the word "ARISE"

HEY LITTLE GIRL THE GULF MY GRANDMOTHER'S HAT WAR IS THE
HIGHEST FORM OF MODERN ART NIGHT IN BAGHDAD THE CARDINAL
POINTS TILT ALIEN SEX THE RIGHT TO BEAR ARMS SPEAK MY
LANGUAGE WHERE I COME FROM ETHEREAL WORMS TIGHTROPE
THE MIND IS A (WILD WHITE HORSE) LA VIDA THAT LITTLE CLOCK

You know the reason why some nights
you don't have a dream?
When there's just blackness?
And total silence?

Well, this is the reason:
It's because on that night
you are in somebody else's dream.
And this is the reason you can't
be in your own dream because
you're already busy
in somebody else's dream.

[221]

In 1992, I was invited to perform in Israel. As an official guest of the government, I met many artists and politicians and did a lot of press conferences. The journalists always began by asking about the avant garde.

One journalist asked, "So what's so good about new?"

"Well," I said, "...uh...um...new is interesting."

"And what," she said, "is so good about 'interesting'?"

"Well, interesting is, you know, interesting...it's uh...like being awake." I was treading water.

"And what's so good about being awake?" she asked. By now I was getting the hang of this: Never actually answer a question in Israel. Always just ask another question.

The Israelis were also very curious about the Gulf War and what Americans had thought about it and I tried to think of a good question to ask in answer to this but what was really on my mind was that the week before I had been testing explosives in a parking lot in Tel Aviv. This happened because I had brought some small bombs to Israel as props for the performance and the Israeli promoter was very interested in them. It turned out that he did weekend duty on one of the bomb squads and bombs were also something of a hobby during the week. He was eager to talk pyrotechnics.

So I said, "Look, these bombs are no big deal, just a little smoke," and he said, "Well, we can get much better stuff for you," and I said, "No, really, these are fine," and he said, "But they should be big, theatrical! I mean you need just the right bombs."

And so one morning he arranged to have about fifty small bombs delivered to a parking lot, and since he looked on it as sort of a special surprise favor, I couldn't really refuse, so we're out in this parking lot, testing the bombs and after the first few I found that I was really getting pretty interested.

They all had very different characteristics. Some erupted in fiery orange sparks and made a low popping sound; others exploded in midair and left long smoky trails. He had several of each kind in case I needed to review them all at the end, and I'm thinking, "Here I am, a citizen of the world's largest arms supplier, setting bombs off with the world's second largest arms customer, and I'm having a great time."

So even though the diplomatic part of the trip wasn't going so well, at least I was getting some instruction in terrorism. And it reminded me of something in a book by Don DeLillo about how terrorists are the only true avant-garde artists *because they're the only ones who are still capable of really surprising people.*

And Jerusalem? It looked just the way my grandmother had described it,
pristine, white, majestic. Except that it was full of guns. Guns and bones.
I'd come to Jerusalem hoping to find out something about time and timelessness,
something about how an Ark could turn into a whale or into a book. But
"Stories from the Nerve Bible" is about the future, and it didn't have any
answers, only questions.

"Come here little girl, get into the car.
It's a brand new Cadillac. Bright red.
Come here little girl."

You know that little clock, the one on your VCR,
the one that's always blinking twelve
because you never figured out how to get in there
And change it.
So it's always the same time
Just the way it came from the factory.
Good morning. Good night.
Same time tomorrow. We're in record.

So here are the questions. Is time long or is it wide?
And the answers?
Sometimes the answers just come in the mail
and you get a letter that says all the things
you were waiting to hear, the things you
suspected, the things you knew were true.
And then in the last line it says:
Burn this.
I think that history has nothing to do
With parades or lines or trains
Or roads. Maybe it's a light
And in that little book called "Einstein's Dreams"
He plays the violin and time stops.
World without end. Remember me.

from "That Little Clock"

[223]

The Right to Bear Arms

You know, in the last year there have been a lot of arguments in my country about the Bill of Rights. Of course, the Bill of Rights is amazing. It guarantees freedom of speech, freedom of the press, religion, the right to trial by jury, and so on but at the moment a lot of Americans are fighting about what the thing really means.

For example, let's take a look at the second amendment, the right to bear arms. This amendment was written two hundred years ago back when people were bagging possum.

So it's pretty hopelessly outdated now as a concept.

The founding fathers also probably hadn't imagined that eighth graders would be showing up at school armed with semi-automatic assault weapons.

And they probably didn't predict that at the end of the twentieth century the privately owned arsenals in this country would dwarf almost every other country's national stockpiles. Because, ladies and gentlemen, the current figures are these: two hundred and sixty million Americans—two hundred million guns.

> Sometimes when you hear someone screaming,
> It goes in one ear and out the other.
> Sometimes when you hear someone screaming,
> It goes right into the middle of your head
> And stays there forever.
> I hear these voices, at the back of my head.
> I'm holding my ears, then my ears turn red.
> My hands are clean. My hands are clean.

from "Where I Come From"

War Is the Highest Form of Modern Art

During the Gulf War I was traveling around Europe with a lot of equipment and all the airports were full of security guards who would suddenly point to a suitcase and start yelling, "WHOSEBAGISTHIS?"

(Explosion)

"I want to know right now who owns this bag!" and huge groups of passengers would start fanning out from the bag just running around in circles like a Scud missile was on its way in.

I was carrying a lot of electronics, so I had to keep unpacking everything and plugging it in and demonstrating how it all worked and I guess it did seem a little fishy, a lot of this stuff wakes up displaying LED-program read-outs that have names like "Atom Smasher" and so it took a while to convince them that they weren't some kind of portable espionage system.

So I've done quite a few of these sort of impromptu new music concerts for small groups of detectives and customs agents. I'd have to set all this stuff up and they'd listen for a while and then say "So what's this?" and I'd pull out something like this filter and say "This is what I like to think of as the voice of authority" and it would take me a while to tell them how I used it for songs that were about various forms of control and they would say:

"Now why would you want to talk like that?"

And I looked around at the swat teams and undercover agents and dogs and the radio in the corner tuned to the Super Bowl coverage of the war and I'd say:

Take a wild guess.

Finally of course I got through. It was after all American-made equipment and the customs agents were all talking about the effectiveness, no, the beauty, the elegance of the American strategy of pinpoint bombing, the high-tech surgical approach which was bring reported on CNN as something between grand opera and the Super Bowl, like the first reports before the blackout when TV was live and everything was heightened, and it was so euphoric.

Night in Baghdad

Oh it's so beautiful, it's like the Fourth of July,
It's like a Christmas tree, it's like fireflies
On a summer night.

Here I'm just going to stick this microphone
Out the window and see if we can't hear a little better.
Can you hear it? Hello, California. Can you hear us?
Come in.

Oh it's so beautiful, it's like the Fourth of July,
It's like a Christmas tree, it's like fireflies
On a summer night.

And I wish I could describe this to you better
But I can't very well right now 'cause I've
Got this damned gas mask on.

So I'm just going to stick this microphone
Out the window and see if we can't hear a little better.
Hello, California? What's the weather like out there?

And I only have one question:
Did you ever really love me?
Only when we danced.
And it was so beautiful. It was like the Fourth of July.
It was like fireflies on a summer night.

Mamacitas, Caballeros,
Welcome to our city, the land of the free.
Welcome to the future,
Hablamos ingles aqui.
Hop in. We're gonna take you for a little ride,
We're goin' downtown.
La vida es un sueño.
Close your eyes now, we're gonna take you there.

I remember where I came from
There were tropical breezes and a wide open sea
I remember my childhood
I remember being free.

La vida es un sueño
Close your eyes now, we're gonna take you there.
La vida es un sueño
Close your eyes now, we're almost there.

Some say the future is written down in a book
And that you can't change a word of it.
And down in the street
They're talkin' on their radios
Gonna build a city that will never fall down.

I remember where I came from
There were burning buildings and a fiery red sea.
I remember all my lovers
I remember how they held me.

Hey, Mister Sandman.
Life is like a dream
But I remember it like a movie.

OK, amigo! Out of the car!
When my father died
We put him in the ground.
When my father died it was like
A whole library had burned down.

from "La Vida" 1992   [227]

Tilt

I find that a good place to think about the future is in airplanes and
airports because time and place start to merge there and everything's in
this constant unstoppable motion. Like the way things have seemed to

speed up in the last couple of years, beginning with the fall of the Berlin Wall. And I remember we watched this on TV every night, over and over, and then one night I thought:

"Gee there's something really familiar about the expressions on their faces."

Then I realized that these were people desperate to shop. They couldn't wait. And somehow from the other side of the wall you got this eerie feeling that you knew exactly what all of this was going to lead to, but they were too far away to scream "GO BACK! GO BACK!" and you didn't know whether it was too early or too late to warn them and anyway they all looked so hopeful.

And ever since the wall fell, it seems like half the world has been pouring from one side to the other, through the train stations, the autobahns, the airports, moving back and forth across the old borders. Like the world had suddenly tilted on its axis and was pouring people from one side to the other. Like an enormous plane, tilting and banking, looking for somewhere to land.

The strangest thing about performing "Stories from the Nerve Bible" in Israel was the show in Tel Aviv. On the screen, there were pictures of buildings that had been blown up in the Gulf War. These buildings had been only blocks away from the theater.

In "Stories from the Nerve Bible" I wore a pair of dark glasses with a tiny video camera attached to the side. During a taped monologue by Admiral Stockdale (an excerpt from the 1992 vice presidential debates), bright lights were turned on in the audience. The camera scanned the first several rows and projected the audience onto the screen.

(On tape, the voice of Admiral Stockdale)

"You know, I ran a civilization of three to four hundred wonderful men. We had our own laws. We had practically our own constitution. And I was the sovereign for a good bit of that. And I tried to analyze human predicaments in that microcosm…in the…in the world.

YOU'VE GOT TO HAVE LEADERS!

And they're out there…who can do this with their bare hands…working with people on the scene.

(Sound of helicopters)
Ladies and gentlemen, there will be four helicopters and the third one
will be Marine One and the President of the United States.

Wild White Horses

In the Tibetan map of the world, the world is a circle and at the center
there is an enormous mountain guarded by four gates. And when they
draw a map of the world, they draw the map in sand, and it takes
months and then when the map is finished, they erase it and throw the
sand into the nearest river.

Last fall the Dalai Lama came to New York City to do a two-week
ceremony called the Kalachakra, which is a prayer to heal the earth.
And woven into these prayers were a series of vows that he asked us to
take and before I knew it I had taken a vow to be kind for the rest of
my life. And I walked out of there and I thought: "For the rest of my
life?? What have I done? This is a disaster!"

And I was really worried. Had I promised too much? Not enough?
I was really in a panic. And there were all these monks walking around.
They had come from Tibet for the ceremony and they were walking
around midtown in their new brown shoes and I went up to one of the
monks and said, "Can you come with me to have a cappuccino right
now and talk?" And so we went to this little Italian place. He had never
had coffee before so he kept talking faster and faster and I kept saying,
"Look, I don't know whether I promised too much or too little. Can
you help me please?":

And he was really being practical. He said, "Look, don't limit yourself. Don't be so strict! Open it up!" He said, "The mind is a wild
white horse and when you make a corral for it make sure it's not too
small. And another thing: when your house burns down, just walk
away. And another thing: Keep your eyes open.

"And one more thing: Keep moving. 'Cause it's a long way home."

[229]

Virtual Reality as the Completion of the Enlightenment Project

Simon Penny

New technologies are often heralded by a rhetoric that locates them as futuristic, without history, or at best arising from a scientific-technical lineage quite separate from cultural history.[1] This paper is an attempt to place virtual reality (VR) and its attendant rhetoric squarely within, and as a product of, the philosophical project of the Enlightenment. Central to this critique is the proposition that while VR is technically advanced, like most computer graphics practices it is philosophically retrogressive.

The key figure here is René Descartes, whose thinking set the scientific and technological revolution in motion. In his famous dream of 1619, he was advised that he should find the unity of the sciences in purely rational terms. The irony at the heart of rationalism is that it arose from a dream. Products of his career that are relevant to this discussion include his famous postulation "cogito ergo sum" (I think therefore I am), his mind-body duality, and Cartesian analytic geometry (1637).[2] In his "Treatise on Man and the Formation of the Foetus," he describes animals in purely mechanistic terms. These are key points in the ideology of rationalism.

VR and Cartesian Space

Recently there has been some criticism of the computer-graphic establishment for its endorsement of a (gendered?) Cartesian space.[3] I would be the first to agree that computer-graphic production—as seen in commercial cinema, video games, theme-park rides, and military simulations—is dominated by a late adolescent Western male psyche and worldview; but I'm troubled by what appears to be an unsubstantiated conceptual leap between the critique of gendered VR spaces and criticism of the utilization of the Cartesian grid.

[231]

One can offer any amount of criticism of the rigidifying effect of the rectilinear grid on our architecture, our city environments, even our 3-D modeling programs. But the Cartesian grid is built into our culture and our perception as an integral and structuring part of the rationalist determinism with which we have been inculcated. To propose an alternative to Cartesian space is to propose an alternative to the philosophical and technical legacy of the Enlightenment.

Cartesian space comprises two parts: the motion of three infinite axes intersecting each other at ninety-degree angles and the division of these axes into continuously discrete quantities. Computer technology is a numerical technology: all data is converted into discrete numerical values, and "space" is broken into numerically defined units. There may well be arguments against Cartesian space as a psychic system, but what alternative space is being proposed or imagined? A system of polar space would seem to invest such space with a center and a phallic core. Cartesian space at least does not privilege any part over any other.

Prior to the formulation of analytic geometry, use of the grid was characteristic of the perspective experiments of Dürer and many other Renaissance draftsmen who sought to accurately represent space from a unique, fixed, monocular viewpoint. But it is this privileging of a sole location of viewing and power in the production of a two-dimensional representation, not the use of a grid, which contains the objectionable theoretical material.

VR and Neovitalism

A criticism levied at computer graphics (at least until *Terminator 2*, *The Lawnmower Man*, and *Jurassic Park*) was that it was a cold space unable to persuasively represent the natural world. This view is equal parts new-age neopagan earthmotherism (enough said), luddism, and narrow-minded critique from the world of traditional art media: it conveniently forgets that the Western tradition took significantly longer to develop these abilities (say, from Cimabue to Leonardo da Vinci). This criticism is now a thing of the past in high-end computer graphics. It remains a problem in VR and in domestic computing. Is the objection to "Cartesian space" voiced as a result of frustration with the inability of current 3-D modeling and world-building kits to produce persuasive amorphous organic forms? This would tally with a somewhat simplistic set of gender-related oppositions (male/female, sharp/soft, square/round, etc.).

On a technical level, the grid (and polygonal construction within the grid) radically limits the possibility of constructing organic, amorphous forms. It privileges clean, crystalline, coherent, independent forms: it is an articulation of the space of the Platonic Ideal. In its abstraction, this space is the space of classical mechanics, of the scientific method, a space in which variables are controlled, systems are assumed to be closed, and nonlinearity has been filtered out. Science lacked mathematical tools to deal with more complex phenomena such as collisions involving more than two bodies, turbulence, and local variation of temperature and pressure. As such, classical mechanics resolve the messy, complex, and overlapping world into clean, self-contained mathematical objects, like the polygonal bodies floating in virtual space.

Resolution, processor speed, bandwidth: these terms are the mantra for computer-graphics engineers. When resolution of the grid moves below the threshold of vision, the problem of organic form will ironically disappear. Ironic, because the increase in resolution that puts the mind at rest is an expression of an increase in power of the technology; however, no intrinsic philosophical change has occurred.

Mimesis in Western Picture-Making

This call for the representation of the "organic" is a call for greater mimesis in computer graphics. But why this obsession with micrographic representationalism? There is an implicit assumption that computer graphics must emulate and exceed the mimetic capability of representational painting and of photography in its representation of organic form according to Renaissance pictorial conventions. The crystalline platonisms may be more "natural," more "native" to that space. Inevi-tably, terms such as "natural," "simulation," and "mimesis" enter the discussion. Historically, mimesis has been a central concern of Western art. Recall the story of a competition between two Greek painters:

Parrhasius entered into a competition with Zeuxis. Zeuxis produced a picture of grapes so dexterously represented that birds began to fly down to eat from the painted vine. Where-upon Parrhasius designed so life-like a picture of a curtain that Zeuxis, proud of the verdict of the birds, requested that the curtain should now be drawn back and the picture displayed. When he realized his mistake, with a modesty that did him honor, he yielded up the palm, saying that whereas he had managed to deceive only birds, Parrhasius had deceived an artist.[4]

This call for a more "organic" representation in the digital realm may be regarded as a retrograde critical position. The most familiar mode of critique of electronic media is characterized by antinomies such as natural/synthetic, authentic/fake, real/artificial—critical systems that ultimately call upon the "natural" as a baseline. But this "baseline" has been well and truly confounded by poststructural critique, which establishes that the "natural" is itself simply a cultural construct. McKenzie Wark has noted that whenever we call upon the organic, the whole, the lying in the past, the lost, we reenact in criticism the myth of the "Fall," the expulsion from the garden of Eden.[5]

VR and Christianity

Brethren, welcome to the First Church of Jesus, Lord of Virtual Space, where you, like Thomas, can thrust your hand in the wound of Christ. You can absolve your sins by taking up His cross on the way to Golgotha. The Church of Jesus, Lord of Virtual Space can deliver to you what the faithful and non-believers alike have yearned for so many generations: Proof!

Here and now I want to extend an invitation to each and every one of you to join me, each Sunday, in the comfort and privacy of your own home, in the most scientific, the most up-to-date religious experience that is available today. Join me every Sunday as we live the life of Christ together. You can witness the Sermon on the Mount, the betrayal in the Garden of Gethsemane: you'll hear the cock crow, you will be there! Don your holy eyephones and come with me, through the chapels of the Stations of the Cross, suffer with your Savior, take Christ into your heart, and step into the body of Christ. Then join me in virtual communion, where angels will pass you golden goblets direct from the flowing wounds of the crucified Christ!

For only $599 and a low monthly rental, you can visit the Kingdom of the Lord (or something very like it) every Sunday. Our Father's house has many mansions, and I want to show you just a few of them.

Picture, if you will, this scene: in living rooms across the country, a parish of believers wearing eyephones, negotiating invisible obstacles, falling on their knees. Inside the eyephones, though, it's a different world: the streets are sandy, the sun Near-Eastern hot, and a crowd is yelling as Pilate hands down his judgment. A Christian virtual theme park. This scenario is an attempt to indicate the likely modes in which technology will manifest itself in a culture such as ours, and also suggests the cultural specificity of *our* VR to a Graeco-Roman tradition, as opposed to a VR that might arise in a non-Western culture, given that such a thing was conceivable.

Virtual reality, like any other technology, is embedded in a cultural history that lends the enterprise a worldview. This worldview is (not unexpectedly) male gendered, patriarchal, and Christian. Jaron Lanier has announced that virtual reality is the culmination of culture—a somewhat conceited judgment, given that he is a major developer of the technology. What concerns me is the cultural specificity of his remark. The abhorrence of the body is inherent in Christian doctrine, and Christianity has served as the basis for Western philosophy until the last century. Philosophical ideas such as the duality of René Descartes are based in Christian doctrine. William Gibson's cyberpunks proclaimed that "the body is meat," but neglected to notice just how similar their position was to that of Saint Augustine. Allucquére Rosanne

Stone has observed that in the Greek New Testament, the word *endyo* is used in the context of narratives of Christian conversion, to mean "to put on Christ" in the sense of putting on an overcoat. In her reading, this condition of "stepping into" is very similar to the condition of being in VR. Such readings strengthen the contention that VR carries with it a philosophical premise rooted in Christian dogma.

VR and Renaissance Humanism

It is a basic tenet of art history that the mode of picture-making practiced in the West is a set of representational conventions devised in the Renaissance. In the computer-graphics community there is a naive acceptance that "Western perspective" is in some sense absolute. This comes as no surprise, since various luminaries have proclaimed that the ultimate goal of computer graphics is to achieve something indistinguishable from a color photograph. Western perspective or, for that matter, any system of pictorial representation is in no sense innate or "natural"; it is an (often arduously) learned convention.[6] The developers of VR have inherited a humanistic worldview (an attitude to life *and* a way of making pictures) that places the eye of the viewer in a position of command, a privileged viewpoint on the world. Recent art theory has examined the relationship between this "Western perspective" and the rise of humanism, rationalism, and empiricism.[7] Computer graphics are therefore heavily invested in the possibility of the objective observer. The notion of the "objective observer" outside the field of action is precisely encapsulated in Western perspective. As Florian Rotzer, in his essay "Interaction and Play," has noted, "the model of the external observer …is gradually being superseded, not only in science but also in art."[8]

VR and Colonialism

Technological development has always defined the location of frontiers. Medieval principalities were limited in scale by the speed of communication and the rate at which troops could be deployed. The Atlantic coast of Europe remained the edge of the world (to Europeans) until explorers were liberated from coast-hugging travel by accurate navigational technologies and robust ships. The American West was claimed and held only once the steam locomotive, the telegraph, and the conoidal bullet combined into one technological complex. More recently, the space race advanced as soon as the technology was available.

With geography filled up and the dreams of space colonization less viable every day, the drive to the frontier has collapsed on itself. The space remaining for colonization is the space of technology itself. No longer the tool by which the frontier is defined, the body of technology is now itself under exploration. Back in the early sixties, one of the pioneers of computer graphics, Ivan Sutherland (who built the first head-mounted display in 1968), declared that the goal was to "break the glass and go inside the machine." More recently, Jaron Lanier has said of VR that "the technology goes away, and all that's left is the cultural component." The technology "goes away" because we are inside it.

VR and the Industrial Revolution

No selection process is value-free, by definition. Software projects are shaped by the worldview of their makers; their value systems are (often unknowingly) incorporated into the work. Computer engineering, software engineering, and knowledge engineering are heirs to the tradition of engineering, the quintessential industrial revolution science, concerned with production—efficient production, by means of standardization of objects and categories.[9]

This process of standardization is antithetical to certain creative goals. Many art movements over the last century have attempted to come to terms with the phenomenon of industrial mass production: Constructivism, the Bauhaus, the Futurists, and "multiples" by artists such as Les Levine. Art has yet to come to terms with this economic-industrial phenomenon. In our historical moment, ideas of "standardization" are being questioned, from social policies of multiculturalism to the instantly reprogrammable robotic production line. The engineering worldview is itself under examination. It seems that the value systems reified in computer technology are dangerously unsteady. One might fairly ask if it is possible to build a "postindustrial" aesthetic within such a steam-powered technology.

VR and Twentieth-Century Philosophical Developments

One might assert that the ideas that have shaped this century can be traced to Marx, Freud, Einstein, and Heisenberg. What effect have their ideas had on the development of the computer? Cartesian rationalism, poststructuralism, feminism, and other varieties of postmodern thought? We might ask ourselves, "If the computer as we know it is heir to such a philosophical system, is it possible to imagine a similar machine based on another system, or is this possibility oxymoronic?" Nell Tenhaaf states, "The philosophy of technology…has been articulated from a masculinist perspective in terms that metaphorize and marginalize the feminine. In real social discourse, this claiming of technology has been reinforced by, and has probably encouraged, a male monopoly on technical expertise, diminishing or excluding the historical contributions of women to technological developments."[10] She goes on to assert that this invisibility of the feminine calls for "a radical reconstitution of technology," but we must ask ourselves whether the architecture of the machine and the premises of software engineering themselves are not so encumbered with old philosophical ideas that any such "reconstitution" would amount only to surface decoration. Key ideas of feminism and poststructuralism include questions of gender, of reowning the body, of the voice of the "other" and "minor literatures," of postcoloniality and subaltern theory. These discourses critique the Enlightenment values of the authority of a rational master discourse and the subjugation and rejection of the body.

[238]

In late 1990, when VR had just burst out of the labs into popular culture, I began to examine some of the rhetorical claims concerning "the body" in VR. In my assessment, VR blithely reifies a mind/body split that is essentially patriarchal and a paradigm of viewing that is phallic, colonializing, and panoptic. A case example of the culturally "male" perspective is the standard paradigm of navigation in virtual space.

Simply stated, in this world of unhindered voyeuristic desire, what the eye wants, the eye gets. It is a machine that articulates scopic desire. But what if VR had developed in a culture with a different attitude to the body? Take, for example, this discussion of Indian dance: "The sense of space was wholly different...no long runs or soaring leaps or efforts to transform the stage into a boundless arena, a kind of metaphysical everywhere...but content within the realm of the body, comfortable with dimension and gravity, all ease, all centered." The teacher of this dance technique described the attitude to the body thus: "no sense of elevation or extension...body self-contained...inwardness, inwardness...In Hinduism, there is no beyond."[11]

In VR all is "beyond," and the body is a void. In the corporeal world, the sense of touch requires immediate physical contact with its object, but the eye does not. VR arms the eye, it gives the eye a hand of its own, propelled (or, more accurately, it appears to be propelled) by the gaze itself. The authoritative viewpoint of Renaissance pictorial space is actively empowered—action at a distance. The entire body is propelled by scopic desire.

The Gendered Gaze, Virtuality, and Safe Sex

Erkki Huhtamo has outlined a history of the "penetration shot" from early cinema through various generations of moving-image technology into the "powerful gaze in VR."[12] He notes the preponderance of environments allowing or facilitating visual command in virtual worlds, as well as the familiar panopticon model, wherein the user oversees an expansive space and can "zoom" to any place at high speed in order to take disciplinary action. If navigation in VR is the articulation of the phallic gaze, we might consider what a feminine alternative might be. [239] A recent work by Agnes Hegedus neatly turns this paradigm on its head: in *Handsight*, the viewer holds a position sensor (polhemus) with the image of an eyeball painted on it, in the hand.[13] If in conventional VR the eye can fly and grab, unhindered by the body, in *Handsight* the body helps the helpless eye about the virtual space. Nor is this virtual space a limitless frontier; rather, it is a closely bounded domain where physical boundaries prohibit the illusion of limitlessness. This inversion is experiential: one discovers it through interaction and consideration.

Lynn Hershman, too, in her interactive laserdisc installation *Deep Contact*, installed at the San Francisco Museum of Modern Art in 1990, interrogates the gendering of electronic tools and spaces by offering the user the prospect of libidinal reward if only "he" would "break through the glass [of the monitor] and touch me."

Paul Sermon has also questioned the gendering of electronic spaces in his video telepresence performance *Telematic Dreaming*.[14] This curious work offers the visitor the pleasure of lying on a bed with a real-time interactive image of Paul, who is in another city, watching the scene on a video monitor. The image of Paul would stroke the visitor sitting or lying on the bed, and the visitor might stroke the image of Paul. Like *Deep Contact*, this work explores questions of technological telepresence via mechanisms of the erotic. A nostalgic reenactment of 1960s love-ins in the erotophobic 1990s, made safe for the age of AIDS, virtual sex is the ultimate prophylactic. VR erotica is just one step further in the process by which successive generations of mechanized-image production (photography, stereo-photographs, offset lithography, cinema...) distance (disembody) the erotic from the biological.

VR, Sci-Fi, and Popular Culture

Virtual reality is discussed here as if it exists. At the time of writing, VR in the civilian domain is a rudimentary technology, as anyone who has worn a pair of eyephones can attest. That the technology is advancing rapidly is perhaps less interesting than the fact that virtually all commentators discuss it as if were fully realized. There is a desire in our culture for VR that can be fairly characterized as a yearning. VR has lingered prenatally in sci-fi and the Star Trek Holodeck for a generation or two, but now is being born. It will slip without friction into our culture because our culture has prepared us for it. I have suggested elsewhere that every significant media technological development since the Renaissance has been employed to create theaters of simulation. André Bazin noted midcentury, regarding the cinema, that, "The guiding myth…inspiring the invention of cinema, is the accomplishment of that which dominated in a more or less vague fashion all the techniques of mechanical reproduction of reality in the nineteenth century, from photography to the phonograph, namely an integral realism, a recreation of the world in its own image, an image unburdened by the freedom of interpretation of the artist or the irreversibility of time."[15]

The way for VR has been prepared by (most recently) Disneyland, Hollywood, liposuction, and Nintendo. We are taught to regard our body as an instrument, as apparel; our culture customizes its bodies like it customizes its cars. The body is a representation only, an external appearance, and may be adjusted to suit the taste of the owner. The absolute malleability of the virtual body is different from this only in degree. In early April 1992, daytime TV host Geraldo Rivera underwent liposuction live on TV in front of a studio audience. Gobs of yellow fat were sucked from his buttocks and injected into his lips and around his eyes. The attitude to the surgical customizing of the flesh, "body sculpting" and the designing of the virtual body, both assumes and reinforces a Cartesian duality by restating the body as pure representation. Thus, VR is an easy step because the body is already a representation.

But how real is VR? The cultural underpinnings are already in place to lubricate the general acceptance that VR does adequately represent "reality." The interchangeability of visual consumption with "experience," which we are encouraged by television to believe, has certainly colored expectations of the virtual environment. The modern notion of functions of the automobile, too, has informed the shape of VR. Iggy Pop defined this condition in his song "The Passenger": "he travels under glass…all of it is yours and mine…so let's ride and ride and ride and ride and ride." It's a very limited interactivity: I can travel and observe, but I cannot act and my environment cannot act upon me. A white man driving through Chicago ghettos in a plush rental car on a hot Saturday evening with the AC and stereo on, with tinted windows and the doors automatically locked, is in VR. The paradigm of VR is informed by the automotive experience: it's a powerful gaze monitoring while remaining undetected, infrared night vision. It's a military

[241]

intelligence model—and why should we be surprised about that?

We are taught to believe that you can "experience" the countryside from inside an air-conditioned car traveling at sixty miles per hour. This "belief" prepares us for the condition of VR: virtual *Reality* is as real as a picture of a toothache. A reality in which you can walk through walls with impunity, a reality that has no odor or temperature—this isn't very real. The construction of increasingly complex and expensive interfaces is beside the point: it's the kind of obsessive project that characterizes the activities of engineers in the realm of cultural production. A discussion of what the "bandwidth" of reality is, is folly. Our preparation is cultural: we will accept VR as a representational scheme, no matter what its verisimilitude, in the same way that we accept a map of a city or the pieces on a chessboard.

VR and the Mind-Body Split

I want to explore now, with respect to Cartesian dualism, the perceptual experience of inhabiting the virtual body. When one wears eyephones and earphones, one shuts out the visual and auditory world and replaces it with a representation. This leaves one part of what we might call the sensorial body in the corporeal world, and the other part in the virtual world.[16] To avoid this split body condition, one must simulate all sensory input in a coordinated way. We might call this *total body representation*. It is instructive to examine what this might imply. If in VR I am confronted with a cast-iron chair, a typist's chair, and a lounge, the ability "to sit" is not enough: the sensation of texture must change, I must also be physically supported by some system. This implies a full "force-feedback"[17] suit unimpending and light enough to be unnoticeable by the wearer. Clearly, this is not feasible. Further, in order for a fully simulated representation of the body to be complete, the realm of the kinesthetic, proprioceptive sense and sense of balance must be addressed. The internal body senses must be "represented"—but how can we electronically simulate the sense of a distended stomach? Sense of taste and smell are also absent. When can I eat virtual food and excrete virtual shit? Clearly, no amount of external gadgetry will facilitate total body representation.[18]

[242]

The VR condition is therefore the limited case of a simulated, inter-active, stereoscopic, visual (and occasionally auditory) environment, in which the body is represented only visually. The prospect of a partial but coordinated and articulated representation of the body raises the question of the repercussions (both psychological and cultural) of a dou-ble body. VR replaces the body with two partial bodies: the corporeal body and an (incomplete) electronic "body image." On a bodily level, the VR experience is therefore one of dislocation and dissociation. Simulator sickness testifies to this dislocation: it is the first virtual illness. The body representation of VR fragments the body, giving rise to a powerful eye mounted on a fractured body. One does not take one's body into VR: one leaves it at the door. VR reinforces Cartesian duality by replacing the body with a body image, a creation of mind (for all "objects" in VR are a product of mind). As such, it is a clear continua-tion of the rationalist dream of disembodied mind, part of the long Western tradition of the denial of the body. Augustine is the patron saint of cyberpunks.

In one of the more absurd examples of sci-fi inspired VR rhetoric, Randy Walser and Eric Gulichsen have said,

> In cyberspace there is no need to move about in a body like the one you possess in physical reality. As you conduct more of your life and affairs in cyberspace, your conditioned notion of a unique and immutable body will give way to a far more liberat-ed notion of "body" as something quite disposable and, gener-ally, limiting. You will find that some bodies work best in some conditions while others work best in others...This is a confu-sion: there is no need for a body at all in VR, except for narcis-sistic or gaming purposes. All one requires is an indication of the location of VR effectors with respect to one's virtual view-point. As the entire physical body is represented in VR by a larger and larger array of interface points, the potential diversity of one's image in VR will become more limited. The variety is possible now only because one can put any shape between the image of the glove and one's virtual viewpoint.

Walser and Gulichsen continue: "The ability to radically and compellingly change one's body image is bound to have a deep psychological effect, calling into question just what you consider yourself to be."[19] Indeed!

There is cause for concern here if our sense of self, our sense of place in the world, remains consistent and continuous purely because external reality has a certain continuity to it. I am suggesting that we have no internal continuous self-image, that self-image is volatile and only a stable "reality" enforces a stable self-image by means of a continually active feedback loop. You have remained a person while reading this paper; you have not turned into a pool of red-hot lava or a collection of reptiles. What, then, are the effects of long-term immersion in VR, of adopting alternative bodies, and what are the effects of "padding" in out in out of a variety of bodies in a variety of worlds? Could the Walser and Gulichsen experience induce schizophrenia?

The term "virtual body" has the ring of "subtle body," a term bandied about amongst adherents of Eastern mystical fads of the 1970s. *There is no such thing as the virtual body*—not in the sense in which we inhabit our bodies. We live in our bodies, and our bodies both act upon the world and register the action of the world upon us. The resistance then collapse of frozen snow under the weight of my foot, the squeaking sound of walking in snow, the pain in my face from the cold, and the gluey feeling of my upper lip as my breath condenses into ice on my mustache are all aspects of my bodily experience of winter. Such holistic body response is not available in VR. In true industrial spirit, duties have been specialized, action is performed by certain parts of the body, and cognition is done by others. The meat body becomes only a machine to press the appropriate buttons or to re-aim the viewpoint, driven by a desiring, controlling mind. The body does not feel, it does not register the virtual world. Only the eyes, privileged as the most "accurate" of the senses since the Renaissance, register the virtual world. Indeed, the "virtual world" is constructed, primarily, as a "visual world"—it is incorporeal!

The Conflation of Representation
with Kinesthetics

Jaron Lanier argues that VR sidesteps the process of translation into, and out of, symbolic representation; he calls this "postsymbolic communication." This claim is, in my view, questionable; Lanier argues that "the way you talk to your body doesn't use symbols.[20] Fair enough, but what is then suggested to be a logical corollary doesn't follow: "you can make a cup that someone else can pick up…without ever having to use a picture or the word 'cup'…you create the experiential object 'cup' rather than the symbolic object."[21] But it's not that simple: the cup in VR is a representation, a stereographic image—you can't drink out of it.

The paradoxical nature of "body" in VR is that the cup itself is an incorporeal image, the movements of my moving arm (in coordination with the image of a hand/tool/pointer moving toward the cup) are bodily experiences. Handing someone a virtual cup resolves the mind/body duality *not* because the virtual cup bypasses the symbolic but because the willful action of "passing the cup" is made. Motor action occurs as a result of will, the real arm moves the representation of the cup, the arm is moving both within VR and without, the realm of representation and physiology are conflated.[22]

There *is* a paradigm shift in the VR experience, but it *does not* bypass the symbolic and replace it with an experience indistinguishable from corporeal experience.[23] The VR representation is an interactive stereographic representation, an automation of pictorial representation. The appellation "virtual reality" is unfortunate for it makes the same sort of untenable claims for the technology that the term "artificial intelligence" did for that discipline. I would prefer to discuss VR as a special, augmented case of representation, such that the object is simultaneously a representation and (in a limited way) a kinesthetically experiential phenomenon. VR directly interfaces with the body, bypassing textual language, but it remains a pictorial representation and is thus subject to critical analysis as such. A new critique is required, a way of thinking about the meeting point between the immediate physiological reality of the body as lived in, and culturally specific conventions of representation.

VR and Power

At the now legendary virtual reality panel at Siggraph '90, VR "came out" to a community of twenty-five thousand at the conference. During a question period, I suggested to the panel that, to my knowledge, there had never been a case in the history of the world when a ruling group did not avail itself of the most advanced technology in order to consolidate or expand its power. I asked the panel why they thought virtual reality would be any different. I was not particularly surprised when the question was politely ignored.

Machine tools, including computers, are devices for exercising power over objects and sometimes over people, but the role of the user in VR is essentially submissive. In VR one *submits* to the representation and the limited freedoms it offers: a postmodern capitalist paradise! This submission is analogous to that of the people employed to manufacture the hardware that runs these virtual worlds: in electronic sweatshops in Taiwan, Malaysia, Mexico, and El Salvador, people (primarily women) labor to produce goods they will never consume. Allucquére Rosanne Stone remarked that René Descartes was able to "forget the body: only because he had servants to attend to the needs of his."[24] Similarly, users of VR implicitly exploit the labor of these Third World workers.

Can we interpret in this swing in VR technology from a paradigm of domination to one of submission as a fin-de-siècle malaise, a simultaneous decline in the will to control, and an acceptance of the overarching power of technology? Gilles Deleuze has built an argument on similar terms, in which he outlines a general movement in the twentieth century from societies of overt discipline, such as Foucault has described, to "ultrarapid forms of free floating control." Deleuze poignantly remarks, "Man is no longer man enclosed, but man in debt."[25] He describes a scenario familiar to consumers of dystopian sci-fi: "a city where one would be able to leave one's apartment, one's street, one's neighborhood, thanks to one's…electronic card that raises a given barrier; but the card could just as easily be rejected on a given day or between certain hours; what counts is not the barrier but the computer that tracks each person's position…and effects a universal modulation."[26]

This is not so much the specter of a machine voice decreeing "access denied" as an internalized control based on knowledge of one's current electronic status. And such a vision is not farfetched: campuses around the country are currently instituting a combined debit and ID card (referred to in some places as an "all-in-one" card) with which students may borrow books, do laundry, work in the computer lab, buy food. Such a card will allow almost real-time tracking of each card owner.

In the computerized workplace, real-time surveillance by means of the computer has been a reality for some time. Paranoia aside, your computer is watching you: text workers are monitored in terms of keystrokes per minute, telephone salespeople in terms of calls per hour. If such a vision is possible in the corporeal world, the prospect of real-time surveillance is so much more simply facilitated in VR: not only will the computer know where you are, but what kind of information you are accessing and where your various body parts are at the time. As digital media become ever-more encapsulating, so the possibility of permanent real-time surveillance becomes real. In a time when rapidly expanding access to electronic information also makes possible automated and invisible systems of surveillance, we must note that the point at which the protection of liberty in the electronic terrain becomes an invasion of privacy is hazy indeed.

In terms of corporate economics, VR serves the computer industry very well: it is intuitive (no learning curve, no consumer resistance), and it calls for unlimited computer power, thus fulfilling the industry's need for technological desire. The transference of libidinal desire onto fetish objects offers the promise of ecstasy but never finally consummates it, driving the consumer to the next purchase in an unending coitus interruptus.

We have no reason to delude ourselves that any new technology, as such, promises any sort of sociocultural liberation. History is against us here. We must assume that the forces of large-scale commodity capitalism will attempt to capitalize fully on the phenomenon in terms of financial profit, and that the potential for surveillance and control will be utilized by corporate and state entities. We have a responsibility to develop a critical consciousness of these possibilities, the better to pre-

[247]

pare ourselves, our children, and students to deal with the highly technologized life style of the early decades of the twenty-first century. As they say, the future isn't what it used to be.

Conclusion

Technologies are products of culture. The ideas that have constructed virtual reality are not new but, rather, have deep roots in our culture. Historically, technological development projects have been considered by their developers as noncontinuous with the world of everyday experience. Virtual reality must not be considered in this way, nor should the developers of these environments be encouraged to think in such a way. It is the fabric of everyday culture that lends meaning, and confines to, in these virtual worlds. The developers and their worlds are immersed in, and informed by, contemporary culture.

I hope to have indicated that the notion of VR as a liberation from the mind/body duality is, like most rhetoric of technology, contrary to fact. This is not to say that VR does not have a complex relation with the body. It is precisely this relation, a relation of instrumentalization, of specialization of parts, that shows it to be the technology at the end of the Enlightenment. Meanwhile, the emergence of postmodern and poststructural theory in the late twentieth century, along with nonlinear dynamics and complexity theory in the sciences, suggests that we are in the process of a shift to a new philosophical model.

Give Me a (Break) Beat!
Sampling and Repetition
in Rap Production

Tricia Rose

You see, you misunderstood
A sample is just a tactic,
A portion of my method, a tool.
In fact it's only of importance
When I make it a priority.
And what we sample is loved by the majority.[1]

In rap, sampling remains a tactical priority. More precisely, samplers are
the quintessential rap production tool. Although rappers did not invent
drum machines or sampling, they have revolutionized their use. Prior
to rap music's redefinition of the role samplers play in musical creativity,
samplers were used almost exclusively as time- and money-saving
devices for producers, engineers, and composers. Samplers were used as
short cuts; sometimes a horn section, a bass drum, or background vocals
would be lifted from a recording easily and quickly, limiting the expense
and effort to locate and compensate studio musicians. Although famous
rock musicians have used recognizable samples from other prominent
musicians as part of their album material, for the most part samples were
used to "flesh out" or accent a musical piece, not to build a new one.[2]
In fact, prior to rap, the most desirable use of a sample was to mask the
sample and its origin; to bury its identity. Rap producers have inverted
this logic, using samples as a point of reference, as a means by which the
process of repetition and recontextualization can be highlighted and
privileged.

Samplers are computers that can digitally duplicate any existing
sounds and play them back in any key or pitch, in any order, sequence
and loop them endlessly. They also have a preprogrammed library of

[249]

digital sounds, sounds that have not been "lifted" from other previously recorded materials but may also be arranged in any fashion. Harry Allen explains: "Record the sound of these pages turning as your TV plays the 'One Life to Live' theme in the background. Or record your boss yelling. Or a piece of Kool and the Gang, whatever, for up to 63 seconds. Loop it, so it plays end-on-end forever, or hook the S900 (sampler) up to a keyboard and play whatever you recorded in a scale."[3]

Samplers allow rap musicians to expand on one of rap's earliest and most central musical characteristics: the break beat. Dubbed the "best part of a great record" by Grandmaster Flash, one of rap's pioneering DJs, the break beat is a section where "the band breaks down, the rhythm section is isolated, basically where the bass guitar and drummer take solos."[4] These break beats are points of rupture in their former contexts, points at which the thematic elements of a musical piece are suspended and the underlying rhythms brought center stage. In the early stages of rap, these break beats formed the core of rap DJs' mixing strategies. Playing the turntables like instruments, these DJs extended the most rhythmically compelling elements in a song, creating a new line composed only of the most climactic point in the "original." The effect is a precursor to the way today's rappers use the "looping" capacity on digital samplers.

Working in the Red

To make the noise that characterizes rap's most creative producers and musicians requires approaching sound and sound manipulation in ways that are unconcerned with the intended or standard use of the samplers. Rap producer Eric (Vietnam) Sadler explains:

[250]

> Turn it all the way up so it's totally distorted and pan it over to the right so you really can't even hear it. Pan it over to the right means put the sound only in the right side speaker, and turn it so you can't barely even hear it—it's just like a noise in the side. Now, engineers...they live by certain rules. They're like, "You can't do that. You don't want a distorted sound, it's not right, it's not correct." With Hank (Shocklee) and Chuck

(D) it's like, "Fuck that it's not correct, just do this shit." And engineers won't do it. So if you start engineering yourself and learning these things yourself—[get] the meter goin' like this [he moves his hand into an imaginary red zone] and you hear the shit cracklin', that's the sound we're lookin' for.[5]

Using the machines in ways that have not been intended, by pushing on established boundaries of music engineering, rap producers have developed an art out of recording with the sound meters well into the distortion zone. When necessary, they deliberately *work in the red*. If recording in the red will produce the heavy dark growling sound desired, rap producers record in the red. If a sampler must be detuned in order to produce a sought after low-frequency hum, then the sampler is detuned. Rap musicians are not the only musicians to push on the limits of high-tech inventions.[6] Yet, the decisions they have made and the directions their creative impulses have taken echo Afro-diasporic musical priorities. Rap production resonates with black cultural priorities in the age of digital reproduction.

Volume, density, and quality of low-sound frequencies are critical features in rap production. Caribbean musics, especially Jamaica's talk over and dub, share a number of similarities with rap's sound. Each feature heavily amplified, prominently featured drum and bass guitar tracks. Both insist on privileging repetition as the basis of rhythm and rhythm as the central musical force.[7] As writers Mark Dery and Bob Doerschuk point out, rappers' production philosophy reflects this emphasis on bass and drum sounds: "To preserve the urgency of rap at its rawest, while keeping the doors of innovation open, a different philosophy of production and engineering has had to evolve…a new generation of technicians is defining the art of rap recording. Old habits learned in MOR, hard rock and R&B do not apply. Like rap itself, the new rules are direct: keep it hot, keep the drums up front and boost that bass."[8] Rap producers use particular digital sound machines because of the types of sound they produce, especially in the lower frequencies.

Boosting the bass is not merely a question of loudness—it is a question of the quality of lower-register sounds at high volumes. The Roland TR-808 is a rap drum machine of choice because of its "fat sonic boom," because of the way it processes bass frequencies. Kurtis Blow explains: "The 808 is great because you can *detune* it and get this low-frequency hum. It's a car speaker destroyer. That's what we try to do as rap producers—break car speakers and house speakers and boom boxes. And the 808 does it. It's African music!"[9] Not only have rap producers selected the machines that allow for greater range of low-frequency resonance, they have also forced sound engineers to revise their mixing strategies to accommodate rap's stylistic priorities.[10] Gary Clugston, rap engineer at INS Recording in New York, explains how rap producers arrange sounds, first pushing the drums to the foreground and at the center of the piece and then using effects to manipulate the bass sounds: "If you're using a drum sample in a rock record, you want it to sit in the mix with everything else. In rap, you do whatever you can to make it stand out—by adding effects, EQ, bottom—and make it sound dirty." "So strong is this fixation with the bass," claim Dery and Doerschuk, "that producers and engineers had to adapt their usual mixing formulas to make room for the rumble." Steve Ett, engineer and co-owner of Chung House of Metal, a popular studio among rap's most prominent producers, elaborates:

> I always put that super-loud long sustaining 808 bass drum on
> track 2. I don't put anything on 1 or 3. If you put the bass
> drum on 2 and the snare on 3, the bass drum *leakage* is tremen-
> dous. It's the only bass drum in the world I'll do that
> with...For me, rap is a matter of pumping the shit out of the
> low end. The bass drum is the loudest thing on the record.
> You definitely hear the vocals but they're very low compared to
> the bass drum.[11]

[252] Ett programs the 808 bass drum knowing that it will have to leak in order to get the desired rumble. This leakage means that the bass will take up more space than is "normally" intended and bleed into other deliberately emptied tracks, which gives the bass a heavier, grittier, less fixed sound. In traditional recording techniques, leakage is a problem to be avoided, it means the sounds on the tracks are not clearly separated, therefore making them less fixed in their articulation. Rock and heavy

metal, among other musical genres, have used distortion and other effects that also require manipulation of traditional recording techniques. Like the use of distortion, if rap's desired sounds require leakage, then leakage is a managed part of a process of achieving desired sounds, rather than a problem of losing control of fixed pitches.

Hank Shocklee prefers the E-mu SP-1200 for its versatility and associates the TR-808 more closely with house music, a dance-music cousin of rap.[12] Most important about his description of these machines is his explanation of how each sampler performs the same technical functions in significantly different ways. Each sampler creates a different feel, thus allowing greater articulation of different rhythmic qualities and musical priorities:

> [The 1200] allows you to do everything with a sample. You can cut it off, you can truncate it really tight, you can run a loop in it, you can cut off certain drum pads. The limitation is that it sounds white, because it's rigid. The Akai Linn [MPC-60] allows you to create more of a feel; that's what Teddy Riley uses to get his swing beats. For an R&B producer, the Linn is the best, because it's a slicker machine. For house records, you want to use the TR-808, because it has that charging feel, like a locomotive coming at you. But every rap producer will tell you that the 1200 is still the ultimate drum machine.[13]

Shocklee prefers the 1200 because it allows him greater cutting and splicing mobility, even though the process of cutting on a 1200 is "stiff." His production work employs the "cut" extensively, demonstrating its capacity to suspend and propel time and motion.[14] Sounding "white" is his reductivist shorthand description for the equipment whose technological parameters adhere most stringently to the Western classical legacy of restricted rhythm in composition. Eric Sadler claims that the TR-808 is still very popular in rap production because of a digital pre-programmed drum sound called the 808 drum boom: "It's not like a regular kick drum. It's this big giant basketball that you hear on just about every record now…Boom…Boom. Big and heavy, just like a reggae sort of feel." Sadler adds that the engineering boards themselves are critical to the feel and sound of rap music, to the process of sound reproduction:

> One of the reasons I'm here (in this studio) is because this board here is bullshit. It's old, it's disgusting, a lot of stuff doesn't work, there are fuses out...to get an old sound. The other room, I use that for something else. All sweet and crispy clear, it's like the Star Wars room. This room is the Titanic room.[15]

Sadler's reference to the Titanic studio can be read as an interesting revision of one of the most well-known black folk toasts, "The Titanic." In it, Shine, the black boiler-room operator on the *Titanic*, tries to warn the white passengers that the ship is about to sink, but his warnings go unheeded. The Captain claims that his water pumps will keep back the water, even though Shine can clearly see they are failing to do so. After a number of warnings, Shine finally jumps overboard saving his own life, saying, "your shittin' is good your shittin' is fine, but here's one time you white folks ain't gonna shit on Shine." Sadler's dubbing that studio the Titanic is his way of saying that it is old and obsolete, suggesting that he chooses the faulty, obsolete equipment deliberately because it allows him to construct his own historical, sonic narratives. The latest, slickest equipment in the Star Wars room denies him access to those sounds and that history. In the hip hop version of "The Titanic," Sadler, like Shine, ignores the white man's definition of technical use and value ("these new valves are better"), but in this case, he does so by staying with his ship, by holding on to the equipment that has been deemed obsolete but best suits his needs. Refusing to follow dominant conceptions of the value of new technology against their better judgment, Shine and Sadler "save" their own lives and narratives respectively.

[254]

These samplers, drum machines, and engineering boards are selected and manipulated by rap producers partially because they allow them to manage repetition and rupture with break beats and looping and cutting techniques and because of the *quality* of sounds they reproduce. Shocklee and Sadler's comments are important to this discussion because they illuminate cultural parameters as they are articulated in advanced electronic equipment. The equipment has to be altered to accommodate rap's use of low-frequency sounds, mixing techniques revised to create the arrangements and relationships between drum sounds. And second, they make clear that rap producers actively and aggressively deploy strategies that revise and manipulate musical technologies so that they will articulate black cultural priorities.

Selecting drum samples also involves matters of sonic preference. Rap's heavy use of sampled live soul and funk drummers adds a desired textural dimension uncommon in other genres and that programmed drum machines cannot duplicate. These soul and funk drummers, recorded under very different circumstances, carry performative resonances that cannot be easily recreated. Bill Stephney, co-owner, with Hank Shocklee, of S.O.U.L. Records, explains why rappers favor particular sources for samples: "They [rap producers] hate digital drums. They like their snares to sound as if they've been recorded in a large live room, with natural skins and lots of reverb. They've tried recording with live drums. But you really cannot replicate those sounds. Maybe it's the way engineers mike, maybe it's the lack of baffles in the room. Who knows? But that's why these kids have to go back to the old records."[16] The quality of sound found in these 1960s and '70s soul and funk records is as important to hip hop's sound as the machines that deconstruct and reformulate them.[17] Rap's sample-heavy sound is digitally reproduced but cannot be digitally created. In other words, the sound of a James Brown or Parliament drum kick or bass line and the equipment that processed it then, as well as the equipment that processes it now, are all central to the way a rap record feels; central to rap's sonic force. This is not to say that live drummers are not featured on rap records, many are; neither is this to say that rap producers do not draw on a wide range of genres, including rap's own previously recorded beats and rhymes. For example, rap's sampling excursions into jazz and rock are increasing all the time. Still soul and funk drum kicks—live or recorded—are almost always the musical glue that binds these samples together, giving the likes of Miles Davis, Ron Carter, Louis Armstrong,

[255]

and Roy Ayers a distinct difference and a hip hop frame. For example, A Tribe Called Quest's "Verses from the Abstract" features Ron Carter on bass but the hip hop drum lines completely recontextualize Carter's jazz sound; similar recontextualizations of jazz samples can be found on a host of rap albums, such as Guru's *Jazzamatazz* and Pete Rock and C. L. Smooth's *Mecca and the Soul Brother*.[18]

Sampling, as it is used by many of hip hop's premiere producers, is not merely a shortcut used to "copy" musical passages. If this were so, then producers would spare the legal costs of clearing usage of other artists' recorded material by programming or replaying very similar musical sequences. Furthermore, as Prince Be Softly of P. M. Dawn points out, finding musical samples can be more time-consuming: "Sampling artistry is a very misunderstood form of music. A lot of people still think sampling is thievery but it can take more time to find the right sample than to make up a riff. I'm a songwriter just like Tracy Chapman or Eric B. and Rakim."[19] The decision to adopt samples of live drum sounds involves quality-of-sound issues and a desire to increase the range of sound possibilities. A few years after rap's recording history began, pioneering rap producer and DJ Marley Marl discovered that real drum sounds could be used in place of simulated drum sounds:

> One day in '81 or '82 we was doin' this remix. I wanted to sample a voice from off this song with an Emulator and accidentally, a snare went through. At first I was like, "That's the wrong thing," but the snare was soundin' good. I kept running the track back and hitting the Emulator. Then I looked at the engineer and said, "You know what this means?! I could take any drum sound from any old record, put it in here and get the old drummer sound on some shit. No more of that dull DMX shit." That day I went out and bought a sampler.[20]

For Marley Marl and other rap producers the sampler is a means to an end, not an end in itself. Nor is it necessarily a shortcut to music production, although some rap producers use samplers and samples in uncreative ways. For the most part, sampling, not unlike versioning practices in Caribbean musics, is about paying homage, an invocation of another's voice to help you say what you want to say.[21] It is also a means of archival research, a process of musical and cultural archeology. According to Daddy-O, rap producer for Stetsasonic, "Sampling's not a lazy man's way. We learn a lot from sampling; it's like school for us. When we sample a portion of a song and repeat it over and over we can better understand the matrix of the song. I don't know how they made those old funk and soul records. We don't know how they miked the drums. But we can learn from their records."[22]

In addition, samples are not strung together in a linear fashion one after the other and then looped. Instead, as Bill Stephney points out, numerous tracks are often programmed simultaneously, sampled on top of one another to create a dense multilayered effect: "These kids will have six tracks of drum programs all at the same time. This is where sampling gets kind of crazy. You may get a kid who puts a kick from one record on one track, a kick from another record on another track, a Linn kick on a third track, and a TR-808 kick on a fourth—all to make one kick!"[23] Once constructed, these looped beats are not set in stone, they are merged with lyrics and reconstructed among other beats, sounds, and melodies. Sadler describes the architectural blueprint for Ice Cube's dense and edgy "The Nigger You Love to Hate":

> The original loop [for it] was [from] Steve Arington's "Weak in the Knees." It was funky but basic all the way through. Cube heard it, liked it, put his vocals on it....Then we stripped it apart like a car and put it back together totally again...erasing musical parts under the choruses and other parts. Everytime we got back to the original song, it would drop down. So we would have to build. That's why the song just kept going up. We kept having to find other parts.

[257]

One of the most dense and cacophonous raps to date, "Night of the Living Baseheads," used nearly forty-five different samples in addition to the basic rhythm tracks and original music on twenty-four tracks. Sadler explains: "Not 48 tracks [which is common in music production today], but 24. You got stuff darting in and out absolutely everywhere. It's like somebody throwing rice at you. You have to grab every little piece and put it in the right place like in a puzzle. Very complicated. All those little snippets and pieces that go in, along with the regular drums that you gotta drop out in order to make room for it."[24]

Rap production involves a wide range of strategies for manipulating rhythm bass frequencies, repetition, and musical breaks. Rap's engineering and mixing strategies address the ways to manage and prioritize high-volume and low-frequency sounds. Selected samplers carry preferred "sonic booms" and aid rap producers in setting multiple rhythmic forces in motion and in recontextualizing and highlighting break beats. These strategies for achieving desired sounds are not random stylistic effects, they are manifestations of approaches to time, motion, and repetition found in many New World black cultural expressions.

Notes

Simon Penny

1. This is the most recent in a series of papers I have written dealing with the cultural context and philosophical implications of virtual reality. My previous work in this field includes: "2000 Years of Virtual Reality," in *Through the Looking Glass: Artists' First Encounters with Virtual Reality*, exhibition catalogue, ed. Janine Cirincione and Brian D'Amato (New York: Jack Tilton Gallery, 1992); and "Virtual Bodybuilding," in *Media Information Australia* (August 1993).

At its outset, VR boggled and challenged many in the techno-arts community, myself included. I have, however, come to the opinion that its past achievements have been mostly technical, rhetoric notwithstanding. The aesthetically difficult work—what to put in virtual worlds, the ways in which these worlds might be constrained and conformed—is yet to be done.

2. Many of these ideas were "of their time": analytic geometry, with the device of axes arranged at ninety degrees with respect to each other, degrees to locate points on the plane or in space, was also developed by Pierre de Fermat around the same time. Fermat's work, however, was not published until 1670.

3. The nano-sex and Virtual Seduction panel at Siggraph '93 was one such case.

4. Quoted in Stephen Bann, *The True Vine: On Visual Representation and the Western Tradition* (New York: Cambridge University Press, 1989), p. 27.

5. Wark raised this notion of retrograde critique at TISEA in Sydney, Australia, 1992.

6. What if VR had developed along pictorial principles other than Renaissance humanism? Could we feel that we might inhabit it at all? In other words, how much is any so-called VR dependent upon culturally acquired knowledge in order to be decipherable? Literary, psychological, and anecdotal evidence attests to the cultural specificity of our particular way of pictorially representing space and distance, relative scale, and so on.

7. Jonathan Crary, among others, discusses these issues in his book *Techniques of the Observer* (Cambridge, Mass.: M.I.T. Press, 1992).

8. Florian Rotzer, "Interaction and Play," in *Machine Culture*, ed. Simon Penny, Siggraph '93, ACM Computer Graphics, special issue.

9. See Manuel De Landa, *War in the Age of Intelligent Machines* (New York: Zone Books, 1991).

10. See *Machine Culture*. Originally published in *Parallelogram* 18, no. 3 (1992).

11. Ross Wetzsteon, "The Cosmic Dance," *Village Voice* 11 February 1992.

12. Erkki Huhtamo, Virtual Zone Symposium, Turku, Finland, October 1992 (text in Finnish).

13. "Ars Electronica," Linz, Austria, June 1992.

14. Performed in Helsinki, Finland (1991).

15. André Bazin, "The Myth of Total Cinema," in *What Is Cinema?* (Berkeley: University of California Press, 1967), p. 21.

16. Can this "body without organs" cope with such fragmentation? The mind seems to willingly close down sensory channels at odds with other more dominant channels; there do seem to be problems when the closed-down channels are being reactivated. Simulator sickness arises from disconnected sensory modalities. "Sitting still" in a flight simulator, in which the image material presents the visual experience of "rolling," requires the mind to prefer the visual input and *ignore* the kinesthetic information from the semicircular canals. This the mind happily does, but it takes at least twenty-four hours for the sense of balance to "reconnect." During that time people tend to fall over; as a result, the navy prohibits the piloting of a plane within twenty-four hours of flight stimulator training.

In 1896–97, G. M. Stratton devised an experiment to invert the visual field by means of an inverting lens placed before the eyes. In an eight-day experiment, he found that by the fifth day he could move about the house with ease. His experiment suggests, to use a gross analogy, that the entire optic nerve bundle was repatched. He notes, however, that when the lenses were removed, the scene had a bewildering quality; there was no sensation that the field was upside down. Perhaps, then, the mind develops a "conditional" model of the world under such circumstances, but readily reverts to the "normal" model. This would be rather like riding a bicycle as opposed to walking. This idea is corroborated by experiments by J. and J. K. Paterson, in which they found that the adaptations made by subjects in order to function with the inverting lenses were instantly recalled when using them again after a lapse of eight months.

Stratton developed another experiment using a mirror arrangement mounted on a shoulder harness which threw his body image out horizontally in front of him. His report—"I had a feeling that I was mentally out of my own body." Similarly, medical records of a patient who had vertebrae removed in surgery reported that he felt that he was viewing the world through his forehead.

People who have a lesion of the parietal lobe in one hemisphere of the brain "have been known to push one of their own legs out of a hospital bed because they were convinced it belonged to a stranger." Such behavior shows that the damaged area normally imparts a signal that says: "This is my body, it is part of myself." (Ronald Melzack, "Phantom Limbs," *Scientific American* [April 1992], p. 123). There is a rare neurological disease that entirely eradicates the kinesthetic sense (the internal sense that allows one to know where one's hands are with one's eyes closed), destroying one's sense of inhabiting one's own body.

Stratton summed up his experiments by saying, "The different sense-perceptions, whatever may be the ultimate course of their extension, are organized into one harmonious spatial system. The harmony is found to consist in having our experiences meet our expectations…The essential conditions of harmony are merely those which are necessary to build up as reliable cross reference between the two senses" (Quoted in R.L. Gregory, *Eye and Brain*, 4th ed. [Princeton: University Press, 1990], p. 214).

Where do these expectations come from? From being in the world, experientially cross-related sense information in early childhood, one supposes. Richard Gregory observes that in order for compensation to take place in a condition of altered sense

information, it is essential that the subject make active corrective movements. He cites the experiments of Richard Held, who finds that active arm movement (striking a target with a finger) is necessary effective adaptation. Gregory asks, "Is the adaptation perceptual or proprioceptive…? (ibid., p. 220). He does not pursue this question, but it is central to the consideration of the cognitive psychology of VR.

The interconnectedness of the "sensorial body" or, as Stratton would have it, the "harmonious spatial system," is born out in another experiment by Held. Two kittens, reared in total darkness, were fitted in an arrangement with two baskets. One basket had holes for the legs such that the physical movements of one kitten would drive both animals through roughly similar spatial experiences. The kitten that could associate visual information with its own physical movement developed effective vision, whereas the other remained unable to coordinate limb movement with visual experience (ibid., p. 219).

These examples have some bearing on the understanding of the body in VR. William Bricken has proposed "superbinocular" vision in VR in which the (virtual) eyes are six feet apart, allowing increased depth perception. In such cases, the existence of a virtual hand would be not simply a useful addition but a necessity if the perceptually different virtual world were to make sense at all. Without the active exploratory hand, this world would remain unintelligible. The construction of a new "sense of body" (even of the limited body of one hand and a pair of eyes) is effected by the physical exploration of this new perceptual space with the hand, and the cross-correlation of internal kinesthetic data with visual input.
17. "Force-feedback" is technology that effects the illusion that virtual objects have some physical mass (put simply: you bump into it, it you can walk through it).

One might argue that the technologies of robotics, space travel, and virtual reality itself, have their roots in literature: robotics in Karel Capek's "RUR" and Fritz Lang's *Metropolis* (these are only two examples), space travel in *Buck Rogers* and the early

Gernsback sci-fi, virtual reality in Disneyland. (These ideas, particularly with respect
to robotics, are discussed in my "The Intelligent Machine as Anti-Christ," SISEA pro-
ceedings, Groningen, Netherlands, 1990).
Another persistent aspect of our sci-fi mythology is the prospect of symbiotic integration
of human organism and machine, I want to
consider some of the implications of the
development of the "direct neural jacks"
that William Gibson has envisaged, from
the perspective of neurology and studies in
perception.

The mind cannot distinguish between
actual heat applied to the hand and the artificial stimulation of nerves someplace between
the hand and the brain. Already nerve interface chips are being developed such that amputees will drive robotic prosthetics through
articulation of the "ghost" limb. So Lanier's
definition of reality as being that which is
"on the other side of the senses" may only
be conditional.

The neuropsychological reality of cognitive remapping is indicated by data from the
applied science of prosthetics. Current technology allows the control of motors in a
prosthetic arm by external electrodes on the
skin adjacent to the muscle mass. This technology is referred to as Myoelectrics. Even
in such hamfisted approaches, the selection
of *which* muscle is somewhat arbitrary: muscles can be retrained such that the mind will
trigger a muscle in the thorax to control the
prosthetic wrist, and this transfer or retrain-
ing becomes "second nature." Performance
artist Stelarc demonstrates this forcefully with
his ability to write simultaneously with three
hands, one of which is a robotic prosthetic
attached at the right elbow. (Stelarc is now
working with a constructible virtual prosthetic for work functions in the virtual
environment.)

Several years ago Bernard Widrow and
his colleagues at Stanford developed experimental implantable microchips, not yet in
the brain, but in the stumps of amputees.
These chips contain fissures through which
severed nerve-ends were encouraged to
grow. The chips contain analogue to digital
converters which would allow the digital

electronic control of servo-motors in robotic
prosthesis (*Chicago Tribune* 9 July 1990).
Experiments on the control of robotic prostheses from nerve impulses from the stumps
of amputees enlarge on this understanding.
There seems no necessity to locate precisely
the nerve for the control of the contraction
of the thumb in order to control a servo-
motor in a robotic prosthesis to control the
robot thumb. Any nerve will do! It might be
the one that controls rotation of the wrist, it
doesn't matter. The mind repatches at *its* end
of the nerve bundle.

How does this neurological research impact the virtual body problem? How readily
will the mind map to the arbitrary virtual
body form? How long will it retain that
mapping as a memory and how many maps
will the mind accommodate simultaneously?
These and related questions remain
unanswered.

The location of the site in the nervous
system or brain where these remapping
operations occur is extremely fugitive.
Recent literature is apt to induce vertigo
as one plumbs a conception of mind that is
incompatible with these gross mechanical
metaphors of plugs and sockets. Ronald
Melzack has recently noted that phantom
limb pain seems not to originate in the
amputee's stump, nor in the spinal chord or
the brain stem, but in a complex parallel
network of higher brain functions. See
Melzack, "Phantom Limbs."

Back in the world of gadget-science,
Thomas Furness relates that the cumbersome
eyephones will be made obsolete by laser
scan direct onto the retina via miniature
hybrid laser chips, which will appear as small
dots in the center of each lens of a pair of
very dark Ray Bans. This image is always in
focus, located at ocular infinity. He predicts
that this technology will be developed in a
few years, and the entire system will weigh
a few ounces and cost $2,000 (Thomas
Furness, Siggraph '91, personal notes).

The prognosis is that the VR interface
gadgets will get smaller and lighter before
they disappear. While they remain attached
to the outside of the body and track actual
physical movement we can continue to

[261]

speak about a dislocation between the virtual body and the corporeal body. But when direct neural jacking synthesizes the entire body experience, the terms of this discussion will radically change.

18. Designer body: As all objects in a virtual world are constructed, so is the body image itself. In "designer reality," the shape and style of the body you take into VR is an open choice. One can design a body with numerous limbs, say a giant lobster, and by attaching additional sensors to knees and elbows to control the extra limbs, one can comfortably inhabit a body with double the regular complement of limbs. The mind maps to this new body almost effortlessly. That is, you begin to instruct your left knee to move, fully knowing that it is in fact the third foot down on the left side. In the case of the Giant Lobster, Lanier reports that it takes only two to three minutes to remap arbitrarily placed sensors as controllers for extra limbs, i.e., sensors on chin or knee. These astonishing reports suggest that the mind can quickly draw a new internal body representation to allow control of the new body. Effectively, to pull the grey matter out of one skull and drop it into another. This effect seems to be at odds with the traditional notion of the neurological homunculus inscribed on the brain. The arbitrary body suggests a way of understanding virtual body articulation as "hyper-marionettry," with the homunculus functioning as a temporary map or I/O program, as opposed to "hardwired" circuitry.

The use of the term "virtual body" is loose and should be clarified. When we discuss the perceiving body, it is in two quite perceptual roles. We can discuss the body as *a thing that is perceived (internally) and understood to be the physical manifestation identified with "self."* In VR this perception is purely visual and it is crudely fashioned. We can also discuss the body as *the thing which does the perceiving of other things outside the body.* In VR this perceiving is specifically visual and auditory.

19. Randy Walser and Eric Gulichsen, quoted in "The Wildest Dreams of Virtual Reality," *M Magazine* (March 1992).

20. In "Revenge of the Nerds," *Afterimage* 18, no. 10 (May 1991), Tim Druckrey interviews Jaron Lanier.

21. Ibid.

22. William Bricken maintains that all the operations of symbolic logic can be performed in VR without recourse to symbolic languages, that logic is equivalent to inference in visual programming. Set theory, number theory, algebra can all be represented as objects in space that is nonsymbolic and totally math-rigorous! Binary logic can be represented as open and shut doors, knot theory as fishing upstream over dams. "All computation is algebraic pattern matching and substitution (proven)" (William Bricken, Siggraph '91, personal notes).

23. Cognition in VR: That VR is incomplete is clear at even a cursory inspection. As a representation, VR is an abstraction. The question is *not* how abstract a representation will the mind/body accept as one which will stand in for reality, as "reality" is an obfuscating word in this context. The question becomes: what constitutes a continuous interactive representation? What arrangement of images and interactive cues cohere into a system with syntactic order? This question is made complex due to the confounding malleability of the mind, what William Bricken refers to as "cognitive remodeling." The mind, it seems, is very willing to restructure itself to compensate for or adapt to a changing "reality." There is a peculiar cognitive feedback loop here: VR, standing in for reality, begins to shape the way the mind describes its experience to itself. The current state of the VR image is extremely simple, built as it is from several thousand polygons. Even so, William Bricken reports that as one interacts with a virtual world, one comes to accept the polygonal representations. It becomes as valid a world as the "real." VR people refer to this as "cognitive plasticity" (William Bricken, Siggraph '91, personal notes). Thomas Furness relates that if you spend a lot of time in VR, you begin to dream in polygons! (Thomas Furness, Siggraph '91, personal notes.) Jaron Lanier's oft-quoted definition that "reality is what is on the

other side of the senses" is validated by these experiences.

The VR representation is ultimately as schematic as that of a map or a chess board. These are schematic representations that are culturally learned, to which we bring meaning and from which we draw meaning. One of the techniques of virtual-world design, as in other computer-interface design, is to utilize the familiar symbols and terminology to indicate to the user that the computer system has been modelled on a familiar "real" world system (i.e., the folders and trash can of the Mac interface). The learning curve is less steep because relationships are familiar. No VR can exist outside of a cultural product.

24. Allucquére Rosanne Stone, quoted in *Telesthesia*, ed. Frances Dyson and Douglas Kahn (San Francisco: San Francisco Art Institute, Walter McBean Gallery, 1991).

25. Gilles Deleuze, "Postscript on the Societies of Control," *October* 59 (Winter 1992), p. 6.

26. Ibid, p. 7. Deleuze is relating a vision of the future originated by Félix Guattari.

Tricia Rose

1. Stetsasonic, "Talk'in All That Jazz," written by Glenn Bolton *In Full Gear* (Tommy Boy, 1988) T-Girl Music Publishing, Inc. (BMI).

2. See Charles Aaron, "Gettin' Paid: Is Sampling Higher Education, or Grand Theft Auto?," *Village Voice Rock 'n' Roll Quarterly* (Fall 1989), pp. 22–23; Jeff Bateman, "Sampling: Sin or Musical Godsend?" *Music Scene* (September–October 1988) pp. 14–15.

3. Harry Allen, "Invisible Band," *Village Voice* 18 October 1988.

4. Phone conversation with Tricia Rose, 14 August 1991. There is quite a large underground market for break-beat records. These LP records are comprised of several rerecorded break beats compiled from other albums. I am aware of at least twenty-five to thirty volumes of these break-beat records.

5. Tricia Rose, interview with Bomb Squad producer Eric (Vietnam) Sadler, 4 September 1991.

6. Decades ago, blues musicians jimmied amplifiers and guitars to get desired sounds, and punk musicians have ignored the official limitations of musical equipment to achieve sought after effects. For other examples, see Kyle Gann, "Sampling: Plundering for Art," *Village Voice* 1 May 1990, p. 102; Andrew Goodwin, "Sample and Hold: Pop Music in the Age of Digital Production," in *On Record: Rock, Pop and the Written Word*, ed. Simon Frith and Andrew Goodwin (New York: Pantheon, 1990), pp. 258–273.

7. See Dick Hebdige, *Cut n Mix: Culture, Identity, and Caribbean Music* (London: Methuen, 1987); especially chapter 10. Rap music is heavily indebted to Jamaican musical practices. Early rap DJs in the Bronx such as DJ Kool Herc were recent Caribbean immigrants and brought with them black Caribbean sound system practices, including sound system wars between DJs. It is also important to stress Jamaican sound systems' emphasis on bass tones. This cross-fertilization is even more complex than immigration patterns suggest. Hebdige demonstrates that reggae's roots are actually in post-World War II black American music. He claims that large powerful sound systems became a popular means by which black American R&B music could be played to large numbers of Jamaicans. See ibid., chapter 7.

8. M. Dery and B. Doerschuk, "Drum Some Kill: The Beat Behind the Rap," *Keyboard* (November 1988), pp. 34–36.

9. Ibid, p. 34. (My italics.)

10. For a discussion of the transformation of the role of recording engineers and their relationship to musicians, see Edward R. Kealy, "From Craft to Art: The Case of Sound Mixers and Popular Music," in *On Record*, ed. Frith and Goodwin, pp. 207–220.

11. Cited in Dery and Doerschuk, "Drum Some Kill," pp. 34–35. (My italics.)

12. Hank Shocklee is a member of the Bomb Squad rap production team which also includes Keith Shocklee, Carl Ryder (Chuck D), and Eric (Vietnam) Sadler. Also, note that house music, a contemporary dance music similar to disco, has been combined with rap to produce Hip House, a popular dance music with rap lyrics.

13. Cited in Mark Dery, "Hank Shocklee: 'Bomb Squad' Leader Declares War on Music," *Keyboard* (September 1990), pp. 82–83, 96.

14. Shocklee's passion for the cut can be best observed in the work of Public Enemy. See especially "Don't Believe the Hype," "Bring the Noise," "Terminator X to the Edge of Panic," and "Night of the Living Baseheads," on *It Takes a Nation of Millions To Hold Us Back* (Def Jam Records, 1988). Similarly, see Eric Sadler DJ Jinx's work on Ice Cube's "The Bomb," *AmeriKKKa's Most Wanted* (Priority Records, 1990).

15. Rose interview with Eric Sadler. For a transcription and interpretation of "The Titanic," see Bruce Jackson, *Get Your Ass In the Water and Swim Like Me* (Cambridge, Mass.: Harvard University Press, 1974). For a provocative reading of the cultural and psychological significance of the sinking of the *Titanic,* particularly as a symbolic representation of the death of civilized European culture, see Mary Anne Doane's reference to Slavoj Žižek in Doane, "Information, Crisis and Catastrophe" in *Logics of Television* ed. Patricia Mellencamp (Bloomington: Indiana University Press, 1990), pp. 229–239.

16. Cited in Dery and Doerschuk, "Drum Some Kill," p. 35. Although Stephney suggests that rappers do not use live drummers with desired success, many albums do feature live drummers in the credits.

17. Sampling attorney Micheline Wolkowicz, who investigates and clears rap samples for Berger, Steingut, Tarnoff, and Stern (a firm that counsels and clears samples for Marley Marl, DJ Jazzy Jeff and the Fresh Prince, the Beastie Boys, and other artists), states that the vast majority of samples cleared by rap musicians are taken from black music performed and created by black musicians. Interview with Rose, September 1991.

18. A Tribe Called Quest, "Verses from the Abstract," *The Low End Theory* (Jive Records, 1991). The title of this album is an obvious affirmation of the importance of low-frequency sounds. Pete Rock and C. L. Smooth, *Mecca and the Soul Brother* (Elektra, 1992); Guru, *Jazzamatazz* (Chrysalis, 1993). See also Ed O. G. and the Bulldogs, "Be a Father to Your Child," *Life of a Kid in the Ghetto* (Polygram Records, 1991).

19. Cited in John Young, "P. M. Dawn Sample Reality," *Musician* (June 1993), p. 23.

20. Cited in Havelock Nelson, "Soul Controller, Sole Survivor," *The Source* (October 1991), p. 38. According to Marley Marl: "'Marley's Scratch' was the first record to use sampled drums, but the innovation really got noticed when it appeared on MC Shan's 'The Bridge' (1986) and Eric B. and Rakim's 'Eric B. Is President' (1986)." Both of these raps were critical successes among hip hop fans and were produced or remixed by Marley Marl.

21. Hebdige, *Cut n Mix*, p. 14.

22. Aaron, "Gettin' Paid," p. 26.

23. Bill Stephney, cited in Dery and Doerschuk, "Drum Some Kill," p. 36.

24. Rose interview with Sadler, 4 September 1991.

V

Technology and the New World Order

Lenin's War, Baudrillard's Games

James Der Derian

> For thirty years, Encore Computer Corporation has envisioned the needs of the most demanding, time-critical flight simulation and training programs in the world. We've understood the pressures for ever-increasing real-time performance; solved complex technical challenges sometimes before our customers knew they had a problem. We've made a business of turning illusion into reality.
>
> —Encore Computer Corporation, advertisement for the 13th Interservice/Industry Training Systems Conference

> The New World Order is messy.
>
> —Robert Murray, "President" of the United States for the 1991 Naval War College Global War Game, Head of the Center for Naval Analyses

Like tracks vanishing in the distance, two parallel memories of terror converge: the memories are of my grandfathers' wars and their work. One grandfather fought a three-year guerilla campaign against the Turks; the other was a machine-gunner in World War I. Neither spoke much of the foreign wars, and I was told not to ask. Their work differed as radically as their wars: one was a straw boss of the coke ovens at the Ford Motor Company in Dearborn, Michigan, the other a bee-keeper in upstate New York.

My first memory invoking them is of a catwalk above the din and heat during my first and last visit to the coke ovens—going into a catatonic terror, having my fingers pried from the railing. My second mem-

[267]

ory is of a Civil War battle, watching my other grandfather, dressed in the uniform of a Union soldier, charge halfway up a hill, discharge a muzzleloader, and die—much too convincingly for a credulous grandson. I grew up intent on keeping my distance from industry and war, real or imaginary. That is until this year, when one empire ended, another went to war, and the repressed returned with a vengeance in the form of an ideological lesson: the end game of simulation in work and war is to make the others' terror fun for us. Needless to say, there is something wrong with this.

A brief historiography of imperialism and simulation might help to explain why. In *Imperialism: The Highest Stage of Capitalism* (1916), Lenin explains why social revolution had not taken place in Europe—contrary to Marx's predictions—but most certainly would in Russia. Lenin assembled a theory of how imperialism would mobilize millions in the belief that their plight was the necessary, final stage of a better world to come. Imperialism, born of the need to keep the working class happy at home, would create a miserable one abroad. Lenin cites the arch-imperialist Cecil Rhodes for his honest appraisal of this fact:

> In order to save the 40,000,000 inhabitants of the United Kingdom from a bloody civil war, we colonial statesmen must acquire new land to settle the surplus population, to provide new markets for the goods produced in the factories and mines. The Empire, as I have always said, is a bread and butter question. If you want to avoid civil war, you must become imperialists.[1]

But as the contradictions of capitalism spread, so too would the formation and solidarity of an oppressed class; world revolution would begin at the periphery and work its way back to the center of capitalist power.

[268]

A fine allegory, Jean Baudrillard would say. Lenin's effort to chart the causes and consequences of imperialism, to pierce the veil of false consciousness that has postponed revolution, to represent scientifically the world to be, merely mirrors, a doubling of the empire's own cartography of the world as it is. "For it is with the same Imperialism," writes Baudrillard, "that present-day simulators try to make the real, all the real coincide with their simulation models."[2] We could go further still, by noting with Nietzsche, that "with the real world we have also abolished the apparent world,"[3] but to save the reality principle, which here means above all the sovereign state acting in an anarchic order to maintain and if possible expand its security and power in the face of penetrating, decentering forces: ICBMs, military and civilian surveillance satellites, international terrorists, telecommunications webs, environmental movements, transnational human rights efforts, to name but a few. In Baudrillard's now familiar words: "It is no longer a question of a false representation of reality (ideology), but of concealing the fact that the real is no longer real."[4]

The idea that reality is blurring or has already disappeared into its representational form has a long lineage. It can be traced from Siegfried Kracauer's chronicling of the emergence of a "cult of distraction" in the Weimar Republic,[5] to Walter Benjamin's incisive warning of the loss of authenticity, aura, and uniqueness in the technical reproduction of reality,[6] to Guy Debord's claim that in modern conditions spectacles accumulate and representations proliferate,[7] and finally, to Baudrillard's own observation that the simulated now precedes and engenders a hyperreality in which origins are forgotten and historical references lost.[8] In his post-Marxist work, Baudrillard describes how the class struggle and the commodity form have dissolved into a universal play of signs, simulacra, and the inertia of mass culture—and, further, how the revolution is missing along with the rest of reality. We are at an end time: but where Lenin saw a relentless, dialectical linearity in capitalism leading to social revolution, Baudrillard sees only a passive population depending on the virtuality of technology to save a defunct reality principle.

Before the last statue of Lenin is toppled and Baudrillard is put in his place, I would like to ask, critically, whether simulation has become the highest stage of technology, and as such, has come to play the fin-de-siècle ideological role once monopolized by imperialism as the last stage of capitalism. War has served as the ultimate test for the ideology of technology of Lenin's imperialism, just as it likely will for Baudrillard's simulation. In Lenin's time trade wars, nationalism, an arms race, the clash and decline of empires, and the inertia of military planning led to a world war so unreal in its level of lethality that, in the end, its only justification was to end all wars. His mechanistic, almost messianic faith is evident in his address to the Third Congress of the Comintern, in which he declared that "the imperialist war of 1914–18 and the Soviet power in Russia are completing the process of converting [the] masses into an active factor in world politics and in the revolutionary destruction of imperialism."[9] Upon signing the Versailles Treaty that ended World War I, President Wilson delivered an equally aspirational message, in the then novel form of a telegram to the American people, describing the treaty as "the charter for a new order of affairs in the world" based on the practice of international law and a promise for international cooperation that would "cleanse the life of the world and facilitate its common action in beneficent service of every kind."[10]

In 1983 Baudrillard noted the dark side of any end to nuclear deterrence:

> Like the real, warfare will no longer have any place—except precisely if the nuclear powers are successful in de-escalation and manage to define new spaces for warfare. If military power, at the cost of de-escalating this marvelously practical madness to the second power, reestablishes a setting for warfare, a confined space that is in fact human, then weapons will regain their use value and their exchange value: it will again be possible to *exchange warfare*.[11]

If ever a war was engendered and preceded by simulation, it was the Gulf War: we were primed for it. Simulations had infiltrated into every area of our lives, in the form of news (re)creations, video games, flight simulators, police interrogations, crime reenactments, and of course, media "war games."[12] Six days into the invasion of Kuwait, Tom Brokaw on NBC news staged a wargame with former U.S. officials standing in for Hussein and Bush. It ended with "Hussein" threatening to "send home body bags every day" and Brokaw warning us that "before too long we may have the real thing." In October Ted Koppel on ABC's "Nightline" weighed in with his Ides of November wargame.[13] Before the ground war, the U.S. conducted a series of highly publicized war exercises, the largest being an amphibious Marine landing called Imminent Thunder. In fact, no landing crafts were used because the seas were running too high. Nonetheless, the simulation "worked." When the allied troops reached Kuwait City they found a room-sized model of the city in a schoolhouse used by the Iraqi military as a headquarters. On a sand tableau there were, to scale, wooden ships, buildings, roads, barbed wire—and all the Iraqi guns pointing toward the sea attack that never came.

The technical reproduction of the war enhanced its hyperreal quality. From the night bombing of Baghdad to the night liberation of Kuwait City, reality was reproduced by night-vision technology and transmitted in real time by portable satellite linkups. The grainy, ghostly green images of the beginning and the end of the war stick: they seem more real, more authentic than all the packaged images sandwiched between them. Call it the new *video verte*: a powerful combination of the latest technology and the lowest-quality image. It reproduced a relative truth: light—a tracer bullet, a secondary explosion, a flaring match—is dangerous, whereas darkness—camouflage, stealth, night— is safety. Correspondents quickly learned that in wartime it was better to dwell and deal in the latter. The motto "We own the night," which originated in the 7th Infantry Division, became the slogan of the war and the reality of its coverage. When obfuscating military briefers and mandatory security reviews extended military hegemony beyond the battlefield, the press and the public, already blind-sided in Grenada and neutered by the pool system in Panama, eagerly seized on the hi-tech prosthetics offered by the military. Words became filler between images produced by gun-cameras using night-vision or infrared that cut through the darkness to find and destroy targets lit up by lasers or radiating heat. Perhaps if a few journalists had known what all night-fighters know, that night-vision degrades depth perception, then the appeal of the videographic reproduction of the war might have been diminished.

Via satellite linkup, ABC correspondent Cokie Roberts posed the question to General Schwarzkopf:

Roberts: You see a building in a sight, it looks more like a video game than anything else. Is there any sort of danger that we don't have any sense of the horrors of war—that it's all a game?

Schwarzkopf: You didn't see me treating it like a game. And you didn't see me laughing and joking while it was going on. There are human lives being lost, and at this stage of the game [sic], this is not a time for frivolity on the part of anybody.

In the space of a single soundbite, Schwarzkopf reveals the inability of the military and the public to maintain the distinction between warring and gaming in the age of video. Many of us were enchanted by the magic of applied technologies, seduced and numbed by the arcane language of the military briefers, satisfied by the image of every bomb finding its predestined target. The wizards in desert khaki came out from behind the curtain only long enough to prove their claims on TV screens, to have us follow their fingers and the arcs of the bombs to the truth. At some moments—the most powerful ones—the link between sign and signifier went into Möbius-strip contortions, as when we saw what the nose of a smart bomb saw as it rode a laser beam to its target, making its fundamental truth-claim not in a flash of illumination but in the emptiness of a dark screen. William Tecumseh Sherman meets Jean-Paul Sartre in a sick syllogism: since war is hell and hell is others, bomb the others into nothingness.

From the initial deployment of troops to the daily order of battle, from the highest reaches of policy making to the lowest levels of field tactics and supply, a series of simulations made the killing more efficient, more unreal, more acceptable.[14] Computer-simulated by private contractors, flight-tested at the Nellis Air Force Base, field-exercised at Fort Irwin in the Mojave Desert, and replayed and fine-tuned every day in the Persian Gulf, real-time wargames took on a life of their own as the real war took lives of countless Iraqis. But there is also evidence that simulations played a critical role in the decision to go to war. Iraq had previously purchased a wargame from the Washington military-consulting firm BDM International to use in its war against Iran; and it was reported in September 1990 on "Nightline" that the software for the Kuwait invasion simulation was also purchased from a U.S. firm.[15] Moreover, Schwarzkopf stated that he programmed "possible conflicts with Iraq on computers almost daily." Having previously served in Tampa, Florida, as head of the U.S. Central Command (at the time a "paper" army without troops, tanks, or aircraft of its own), his affinity for simulations is unsurprising.

In fact, Schwarzkopf sponsored a highly significant computer-simulated command-post exercise which was played in July 1990 under the code name of Exercise Internal Look '90. According to a Central Command news release issued at the time, "command and control elements from all branches of the military will be responding to real-world scenarios similar to those they might be expected to confront within the Central Command AOR consisting of the Horn of Africa, the Middle East and Southwest Asia." The wargame specialist who put Exercise Internal Look together, Lieutenant General Yeosock, moved from fighting "real-world scenarios" in Florida to commanding all ground troops except for the special forces under Schwarzkopf.

Perhaps it is too absurd to believe that the Gulf War is in some sense the product of a wargame designed to fight another wargame, perhaps not. My purpose is not to conduct an internal critique of the simulation industry, nor to claim some privileged ground that permits me to discover the causes of the war.[16] Rather, my intent is to ask if, in the construction of a realm of meaning that had minimal contact to historically specific events or actors, simulations demonstrated the power to construct the reality they purport to represent. The question is whether simulations can create a New World Order in which actors act and things happen that have profound consequences outside of the artificial cyberspace of the simulations themselves.

Over the last decade there has been a profusion of signs that a simulation syndrome has taken hold in international politics. According to Oleg Gordievsky, former KGB station chief in London, the Soviet leadership became convinced in November 1983 that a NATO command post simulation called Able Archer '83 was in fact the first step toward a nuclear surprise attack.[17] Relations were already tense after the

September shootdown of the Kuwaiti Air Lines flight 007—a flight that the Soviets interpreted as an intelligence-gathering mission—and since the Warsaw Pact had their own wargame using a training exercise as cover for a surprise attack, the Soviets assumed the West to have one as well. No NATO nuclear forces actually went on alert, yet the KGB reported the opposite to Moscow. On November 8 or 9, flash messages were sent to all Soviet embassies in Europe, warning them of NATO preparations for a nuclear first strike. Things calmed down when the NATO exercise ended without the feared nuclear strike, but Gordievsky still maintains that only the Cuban Missile Crisis brought the world closer to the brink of nuclear war.

The Gulf War is the preeminent but probably not the last case of a simulation syndrome manifesting itself in the military and the media. Baudrillard was right in the sense that simulations would rule not only in the war-without-warring of nuclear deterrence but also in the postwar warring of the present.[18] It was never in question that the coalition forces would win the military conflict, but they did not win a "war" in the conventional sense of destroying a reciprocating enemy. What, then, did the U.S. win? A cyberwar of simulations. First, the prewar simulation, Operation Internal Look '90, which defeated the Iraqi simulation for the invasion of Kuwait; second, the wargame of Air-Land battle, which defeated an Iraqi army that resembled the game's intended enemy, the Warsaw Pact, in hyperreality only; third, the war of spectacle, which defeated the spectacle of war on the battlefield of videographic reproduction; and finally, the postwar after-simulation of Vietnam, which defeated an earlier defeat by assimilating Vietnam's history and lessons into the victory of the Gulf War.

There is, of course, a fundamental and ultimate difference between war and its game—people die in wars. But this distinction also suffered representational erosion in the Gulf War. Did the simulation syndrome kick the Vietnam Syndrome? I'm sure that as long as there is a great global gap in power and wealth there will be tenacious underdogs with a taste for gray flannel—and more swift kicks to follow. But the score is being kept. Almost twenty-five years ago at the Bertrand Russell War Crimes Tribunal in Stockholm, Jean-Paul Sartre rendered a verdict that links Lenin's war to Baudrillard's game: the United States, he said,

> is guilty, by plotting, misrepresenting, lying and self-deceiving, of becoming more deeply committed every instant, despite the lessons of this unique and intolerable experience, to a course which is leading it to the point of no return. It is guilty, self-confessedly, of knowingly carrying on this *cautionary* war to make genocide a challenge and a threat to peoples everywhere. We have seen that one of the features of total war was a constant growth in the number and speed of means of transport; since 1914, war can no longer remain localized, it must spread through the world. Today the process is becoming intensified; the links of the *One World*, this universe upon which the United States wishes to impose its hegemony, are ever closer.[19]

Perhaps Lenin's and Baudrillard's worst scenarios have come true: the post–Cold War security state now has the technology of simulation as well as the ideological advantage of unipolarity to regenerate, at relatively low cost to itself, an ailing national economy and identity through foreign adventures. We should expect, then, endo- as well as exo-colonial wars, trade wars, and simulated wars. Iraq served its purpose well as the enemy other that helped to redefine the Western identity. But it was the other enemy, the more pervasive and elusive threat posed by the deterritorialization of the state and the disintegration of a bipolar order that has left us with a Gulf War Syndrome, in which the construction and destruction of the enemy other is measured in time, not territory; prosecuted in the field of perception, not politics; authenticated by technical reproduction, not material referents; and played out in the method and metaphor of gaming, not the history and horror of warring.

The only alternative, as I see it, is to find a home in the new disorder by a commensurate deterritorialization of theory. We can no longer reconstitute a single site of meaning or reconstruct some Marxist or even neo-Kantian cosmopolitan community; that would require a moment of enlightened universal certainty that crumbled long before the

Berlin Wall fell. Nor can we depend on or believe in some spiritual, dialectical, or technological process to overcome or transcend the domestic and international divisions, ambiguities, and uncertainties that mark the age of speed, surveillance, and simulation. Rather, we must find a way to live with and to recognize the very necessity of difference, the need to assert heterogeneity against the homogenizing and often brutalizing force of technology. This is not yet another utopian scheme to take us out of the "real" world but, rather, a practical strategy to mediate rather than exacerbate the anxiety, insecurity, and fear of a New World Order in which radical otherness is ubiquitous and indomitable.

Video/Television/Rodney King: Twelve Steps Beyond the Pleasure Principle

Avital Ronell

CHANNEL TWELVE Ethics has been confined largely to the domains of doing, which include performative acts of a linguistic nature. While we have understood that there is no decision that has not passed through the crucible of undecidability, ethics still engages, in the largest possible terms, a reflection on doing. Now, what about the wasted, condemned bodies that crumble before a television? What kinds of evaluations, political or moral, accrue to the evacuated gleam of one who is wasting time—or wasted by time? There is perhaps little that is more innocent, or more neutral, than the passivity of the telespectator. Yet in *Dispatches*, Michael Herr writes, "it took the war to teach it, that you were as responsible for everything you saw as you were for everything you did. The problem was that you didn't always know what you were seeing until later, maybe years later, that a lot of it never made it in at all, it just stayed stored there in your eyes."[1] What might especially interest us here is the fact that responsibility no longer pivots on a notion of interiority: seeing itself, without the assistance of cognition or memory, suffices to make the subject responsible. It is a responsibility that is neither alert, vigilant, particularly present, nor informed.

HEADLINE NEWS *Testimonial video functions as the* objet petit *for justice and the legal system, within which it marks a redundancy, and of which it is the remainder.*

[277]

CHANNEL ELEVEN The lawyers defending the four policemen approached George Holiday's videotape by replicating the violence that had been done to Mr. Rodney King.[2] The unquestioned premise upon which the team of lawyers based their defense of the police called for an interpretation of video in terms of a frame-by-frame procedure. No one questioned this act of framing, and the ensuing verdict unleashed the violence that would explode the frames set up by the court. In the blow-by-blow account, counting and recounting the event of the beating, the defense presented a slow-motion sequencing of photographs whose rhythm of articulation beat a scratchless track into the court records. The decisive moves that were made on video require us to review the way in which media technology inflects decisions of state. That would be the larger picture. The smaller picture, encapsulated by the larger one, concerns the legal ramifications of distinct interpretive maneuvers. Thus, the chilling effects of warping video into freeze-frame photography cannot be overlooked—even where *overlooking* can be said to characterize the predicament in which testimonial video places the law. For the duration of the trial, the temporization that reading video customarily entails was halted by spatial determinations that were bound to refigure the violence to which Mr. King was submitted. No one needs to read Jacques Derrida's work on framing in order to know that justice was not served in Simi Valley, California. But, possibly, if one had concerned oneself with the entire problem of the frame, its installation and effects of violence—indeed, the *excessive force* that acts of framing always imply—then it would have been imperative to understand what it means to convert in a court of law a videotape into a photograph. For the photograph, according to the works of Walter Benjamin, Roland Barthes, Jacques Derrida, and a number of others, draws upon phantasmal anxieties as well as the subject's inexorable arrest. I need not stress to what extent the black body in the history of racist phantasms has been associated with the ghost or zombie.

Perhaps we ought to begin, then, with the astonishing remark of Jacques Lacan when he was on television, later translated in *October*:

—*From another direction, what gives you the confidence to prophesy the rise of racism? And why the devil do you have to speak of it?*
—Because it doesn't strike me as funny and yet, it's true. With our *jouissance* going off the track, only the Other is able to mark its position, but only insofar as we are separated from this Other. Whence certain fantasies—unheard of before the melting pot.

Leaving this Other to his own mode of *jouissance*, that would only be possible by not imposing our own on him, by not thinking of him as underdeveloped.[3]

It would appear that, in "Television," the incompletion of our *jouissance* is marked with some measure of clarity only by the Other; or, at least its off-track predicament engages a boundary that exposes the Other to the projections of racist fantasies. Claiming the relative stability of a position, the Other becomes the place that failed *jouissance* targets, if only because it provides a range of separation. This separation, this tele-, constitutes the distance we have to travel, whether this be accomplished handheld or alone, on the streets or *in camera*. Of the fantasies that set off the signals for mutilating the body of Rodney King, one involved precisely a kind of tele-vision that could see little more than the *jouissance* of the Other, a night vision flashing a second degree of self that emerges with destructive *jouissance*: the night blindness that operates the intricate network responsible for the policing of drugs.

In order to get in gear, the police force had to imagine their suspect on PCP, and they fantasized, they claimed, that they were considerably threatened by the solitary figure. What does it mean to say that the police force is hallucinating drugs, or, in this case, to allow that it was already in the projection booth as concerns Rodney King? In—before, in fact—the first place, they were watching the phantom of racist footage. According to black-and-white TV, Rodney King could not be merely by himself or who he was that night. In order to break Rodney King, or break the story, the phantasm of the Other, supplemented on junk—beside himself, not himself, more than himself, a technozombie of supernatural capabilities—had to be agreed upon by the police force. The police reached such a consensus on location. So (we are reconstructing the politico-topography slowly), the Rodney King event was articulated—in the first place—as a metonymy of the war on drugs; this war, ever displacing its target zones, licensing acts of ethnocide on the part of hallucinating mainstreamers.

But there are other places and other types of projections that come to light here: the Rodney King event is equally that which opens the dossier of the effaced Gulf War. When television collapses into a blank stare, whiting out the Gulf War, nomadic video in turn flashes a metonymy of police action perpetrated upon a black body. While things and connections should be encouraged to become clear, they should not perhaps hold out expectations of becoming, once and for all, "perfectly clear"—a common idiom for the white lie.

The empirical gesture through which the violence erupted on March 3, 1991, was linked to Rodney King's legs. Did he take a step or was he charging the police? The footage seemed unclear. The defense team charged that King had in fact charged the police. "Gehen wir darum einen Schritt weiter," writes Freud in *Beyond the Pleasure Principle*—a text bringing together the topoi of charges, repetition compulsion, violence, and phantasms. "Let us take another step further," and another, and as many as it takes, in order to read the charges that are electrifying our derelict community.

HEADLINE NEWS Read the step digitally: crime serials/serial murders.

CHANNEL TEN Unlike telephony, cinema, or locomotion, the mass invasion of television emerged as a prominent figure of our time only *after* World War II. There are many reasons for this (for example, the Nazis voted in radio as the transferential agency par excellence, television was canceled out of the secret service of fascisoid transfixion). Served on the Cold War platter, it is not so much the beginning of something new but, instead, the residue of an inassimilable history. Television, linked to the enigma of survival, inhabits the contiguous neighborhoods of broken experience and rerouted memory. Refusing in its discourse and values to record, but preferring instead to play out the myths of liveness, living color, and being there, television will have produced a counterphobic perspective to an *interrupted history*. I hope to scan the way TV acts as a shock absorber to the incomprehension of survival, being "live" or outliving as the critical enigma of our time.

[280]

Walter Benjamin theorized the difference between *überleben* and *weiterleben*, surviving and living on. Television plays out the tensions between these modalities of being by producing narratives that compulsively turn around crime. These narratives, traveling between real and fictive reference, allow for no loose ends but suture and resolve the enigmas they name. Television produces corpses that need not be mourned because, in part, of the status of surviving that is shown. Still, television itself is cut up, lacerated, serrated, commercial-broken, so that its heterogeneous corpus can let something other than itself leak out.

CHANNEL NINE The death of God has left us with a lot of appliances. Indeed, the historical event we call the "death of God" is inscribed within the last metaphysical spasm of our history as it continues to be interrogated by the question of technology. The event of the death of God, which dispersed and channeled the sacred according to altogether new protocols, is circuited through much of technology. I refer to God because He in part was the guarantor of absolute representability and the luminous truth of transparency. In an era of constitutive opaqueness—we dwell not in transcendental light but in the shadows of mediation and withdrawal; there will be no revelation, can be no manifestation as such—things have to be tuned in, adjusted, subjected to double-takes and are dominated by amnesia. Without recourse to any dialectic of incarnation, something yet beams through, as though the interruption itself were the thing to watch.

CHANNEL NINE Media technology has made an irreversible incursion into the domain of American "politics." The ensuing anxieties concern not so much the nature of fictioning—politics has always been subject to representation, rhetoric, artifice—as it does newly intrusive effects of law. This is not to say that the law has ever been zoned outside of us but, thanks to the media, different maps of arrest have been drawn up: the subject is being arrested according to altogether new protocols of containment. (Practically everybody in homeless America is under house arrest.) The Rodney King episode is exemplary because it functions as an *example*—something that implies a generality of which it is a part; but, at the same time, one must not lose sight of the singularity it brings to bear upon our understanding of media and state politics.

[281]

Few episodes have broken the assignments of their frames and exploded into the socius as has the Rodney King "event." Beyond the articulated outrage that this episode has produced, its persistent visibility has forced us to ask tough questions about American scenes of violence.

When they pass into the media and graduate into "events," are these scenes already *effects* of the media? To what extent is serial murder an *effect* of serial television or its imprinting upon a national unconscious? Or, to return to our channel: what is the relationship between mutant forms of racism (today's racism differs from yesterday's: it is constituted by a different transmitter) and the media, which appears to resist older types of racism?

One may wonder why it was the case that the Los Angeles Police Department, and not an equally pernicious police force, became the object of coverage by television. On some crucial level, television owns and recognizes itself in the LAPD—think of *Dragnet* and the like producing the mythic dimension of policing. The LAPD is divided by referential effects of historical and televisual narration. That the one is constantly exposed by the other, flashing its badge or serial number, is something we now need to interrogate. Exposed as the LAPD appears to be, it is always covered by television.

But what is television covering? This would seem to bring us a long way from the question of politics in America. Yet, owing to the teletopies created by television when it maps political sites, we no longer know where to locate the polis, much less "home." This is why we must begin with the most relentless of home fronts.

HEADLINE NEWS *Some of you think that it is the hour of TV-guided destitution. Inside and outside the home the time has come to think about the wasted, condemned bodies that crumble before a television.*

CHANNEL EIGHT Among the things that TV has insisted upon, little is more prevalent than the interruption or the hiatus for which it speaks and of which it is a part. Television persists in a permanent state of urgency—hence the necessity of the series. The series, or seriature, extradites television to a mode of reading in which interruption insists, even if it does so as an interrupted discourse whose "aim is to recapture its own rupture."[4] If we are going to attempt to read the interruption as such, then we shall be reading something that is no longer appropriable as a phenomenon of essence. The question remains whether we can read the trace of interruption that is put into some sort of shape by a series or net without it being dominated by the logic of the cut but, rather, by tracking the intricacies of *destricturation*: the interruptions that constitute a network where knots never land or tie up but, rather, cause traces of intervals to be indicated. Precisely such fugitive intervals—as elusive slip or trace of interruption—bind us ethically. Through its singular mode of persistence, the hiatus announces nothing other than the necessity of enchaining the moments, particularly the moments of rupture, albeit in a nondialectical fashion: it would be absurd to make claims for producing a dialectical summation of all the series, that is, interruptions, that tie up TV and its breaks. Where the effect of shock continues to jolt or to make the image jump, there is still a need to enchain the moments and produce linkage.

So, in the space of interruption, between Kansas and Oz (an atopy or interruption that used to be called "television land"), there exists a muted injunction to read the hiatus and let oneself be marked by the hiatus—a necessity of negotiating the invisible lineage of the net.[5] To this end, it becomes necessary to displace the focus from television as totality to the seriality of derangement, a place of disturbance, something that can be designated as being "on location" only on the condition that it remain dislocated, disarticulated, made inadequate and anterior to itself, absolutely primitive with regard to what is said about it. If TV has taught us anything—and I think it is helpful to locate, somewhere between Kansas and Oz, an internal spread of exteriority, an interruption precisely of the phantasmatic difference between interiority and exteriority—the teaching principally concerns, I think, the *impossibility of staying at home*. In fact, the more local it gets, the more uncanny, not-at-home, it appears. Television, which Heidegger when he was on, once associated with the essence of his thinking, chains you and fascinates you by its neutral gleam; it is about being-not-at-home, telling you that you are chained to the deracinating grid of being-in-the-world. Perhaps this explains why, during his broadcast, Lacan spoke of *homme*-sickness.[6] We miss being-at-home in the world, which never happened anyway, and missing home, Lacan suggests, has everything to do in the age of technological dominion with being sick of *homme*. So where were we? asks the scholar.

We have no way of stabilizing or locating with certainty the "in" of being-in-the-world. Television exposes that constitutive outside that you have to let into the house of being, inundating and saturating you, even when it is "off." While television, regardless of its content or signified, tends toward an ontologization of its status—no matter what's on, it is emptied of any signified—it is a site of evacuation, the hemorrhaging of meaning, ever disrupting its semantic fields and the phenomenal activities of showing—television commits a trace to the articulation of sheer uncanniness. This is what Heidegger understands as our fundamental predicament of being-not-at-home in this world (which we have yet to locate and which technology helps to map in terms of teletopies, which is of no help). While television's tendential urge guides it toward ontology, there are internal limits that freeze-frame the ontological urge into an ethical compulsion. One of these internal limits that is at once lodged outside *and* inside TV is a certain type of monitoring—the nomadic, aleatory, unpredictable eruption sometimes called "video."

CHANNEL SEVEN TV has always been under surveillance. From credit attributions to ratings and censorship concessions, television consistently swerves from the ontological tendency to the establishment of legitimacy, which places it under pressure from an entirely other obligation. It is no wonder that television keeps on interrupting itself and replaying to itself the serial crime stories that establish some provisional adjudication between what can be seen and an ethico-legal resolution to the programs of showing. Oedipus has never stopped running through television, but I'll get to the violence of legitimation and patricidal shooting momentarily. The crime stories that TV compulsively tells itself have been charged with possession of a mimetic trigger—in other words, as TV allegorizes its interrupted relation to law, it is charged with producing a contagion of violence. A perpetual matter of dispute, the relation of television to violence is, however, neither contingent nor arbitrary but zooms in on the absent, evacuated center of televisual seriature. At the core of a hiatus that pulses television, I am placing the mutism of video, the strategy of its silence and concealment. Though I recognize the radically different usages to which they have been put empirically, and the divergent syntaxes that govern their behavior, I am interested in the

[284]

interpellation that takes place between television and video, the way the one calls the other to order. In fact, where nomadic or testimonial video practices a strategy of silence, concealment, and unrehearsed semantics, installed as it is in television as bug or parasite, watching (out for) television, it at times produces the ethical scream that television has interrupted.

This ethical scream that interrupts a discourse of effacement (even if that effacement should indeed thematize crime and its legal, moral, or police resolutions), this ethical scream—and video means for us "I saw it"—perforates television from an inner periphery, instituting a break in the compulsive effacement to which television is in fact seriously committed. When testimonial video breaks out of concealment and into the television programming that it occasionally supersedes, it is acting as the call of conscience of television. This is why, also, when television wants to simulate a call of conscience (*Gewissensrufen*, the aphonic call discussed in Heidegger's *Being and Time*), it reverts to video. The abyssal inclusion of video as call of conscience offers no easy transparency but requires a reading; it calls for a discourse.

As we have been shown with singular clarity in the Rodney King case and in particular with the trial, what is called for when video acts as television's call of conscience is not so much a viewing of a spectacle, but a reading and, instead of voyeurism, an exegesis. On both sides of the showing of this video, we are confronted with the image of condemned and deserted bodies, the "trash bodies," dejected and wasted— and this is why, when you're on television, as its spectral subject on either side of the screen, you've been trashed, even if watching television is only a metonymy of being wasted in the form, for instance, of "wasting time." To the extent that we, like Rodney King, are shown being wasted by the deregulation of force and are left crumpled by the wayside, we know we are dealing with a spectral experience and the screen memory of the phantom. At the same time, we must also endeavor to understand why the generative impulse of the Rodney King story pivoted on the misappropriation of drugs—by the police. TV on drugs, policing drugs: the working out of a hallucinogenre in which the suspect is, on the surface, viewed as being on PCP, a problem relating *aletheia* to phantomal force. The collapse of police into television is only beginning to produce a history of phenomenal proportions. As the technical medium returns to the site of its haunted origins, it shows one of the more daunting aspects of the collaboration between law enforcement and performative television to consist in the growing number of arrests made by the latter.[7]

Haunted TV: the phantom of the Gulf War, bleeding through the body of King—showing by not showing what lay at our feet, the step out of line, focusing the limitless figure of the police, this "index of a phantom-like violence because they are everywhere."[8] The police aren't just the police "today more or less than ever," writes Derrida: "they are the *figure sans figure*."[9] They cut a faceless figure, a violence without a form, or, as Walter Benjamin puts it: *gestaltlos*. This formless, ungraspable figure of the police, even as it is metonymized, spectralized, and even if it installs its haunting presence everywhere, remains, for Benjamin, a determinable figure properly belonging to the concept of the civilized state. This is why we are never going to be on furlough as concerns the necessity of reading the effects of haunted TV.

The Rodney King interruption not only forces a reading of force and enforcement of law but also requires citation and a reading precisely of the phantom body of the police. If anything, it announced the ubiquity of the police station identification. The police become hallucinatory and spectral because they haunt everything; they are everywhere, even where they are not. They are present in a way that does not coincide with presence: *they are television*. But when they come after you and beat you, they are like those televisions that explode into a human *Dasein* and break into a heterotopy that stings. Always on, always on your case, always in your face.

[286]   What video teaches, something that television knows but cannot articulate as such, is that every medium is related in some crucial way to specters. This ghostly relationship that the image produces between phenomenal and referential effects of language is what makes ethical phrasing as precarious as it is necessary. Because of its transmission of ghostly figures, interruption, and seriature, we cannot merely assimilate television to the Frankfurt School's subsumption of it under the regime

of the visual, which is associated with mass media and the threat of a fascist culture. This threat always exists.

Television, in its couple with video, offers a picture of numbed resistance to the unlacerated regimes of fascist media as it mutates into forms of video and cybernetic technology, electronic reproduction and cybervisual technologies.

HEADLINE NEWS *The disfigured writing on the face of Rodney King.*

CHANNEL SIX Television is being switched on for various reasons. Despite and beside itself, it has become the atopical locus of the ethical implant. Not when it is itself, if that should ever occur or stabilize, but when it jumps up and down on the static machine, interfering as the alterity of constant disbandment.

One problem with television is that it exists in trauma or, rather, trauma is what preoccupies television: it is always on television. This presents us with considerable technical difficulty, for trauma undermines experience even as it acts as its tremendous retainer. The technical difficulty consists in the fact that trauma can be experienced in at least two ways, both of which block normal channels of transmission: as a memory one cannot integrate into one's own experience, and as a catastrophic knowledge that one cannot communicate. If television can neither be hooked up to what we commonly understand by the term *experience* nor communicate or even telecommunicate a catastrophic knowledge, but *can* only—perhaps—signal the transmission of a gap (at times a yawn), a dark abyss, or the black box of talking survival, then what has it been doing? Why—or how—does it induce the response of non-response and get strapped with charges of violent inducement?

CHANNEL FIVE When adjusting myself to technology I was more pointedly drawn in by the umbilicus of the telephone, whose speculative logic kept me on a rather short leash. On first sight, television seemed like a corruption, as is true of many supercessions, of the serious lineage of telephony; it seemed like a low-grade transferential apparatus, and I felt television and telephone fight it out as in the battle in *Robocop* between mere robot (His Majesty, the ego) and the highly complex cyborg (who came equipped with memory traces, superego, id, and, displacing the ego, a crypt). In yet another idiom, telephony was for me linked to the Old Testament (the polite relation to God, as Nietzsche says), where television seemed like the image-laden New Testament (where one rudely assumes an intimacy with God and makes one of His images appear on the screen of our historical memory). Needless to say, I was on the side of the more remote, less controlled, audible sacred, to the extent that its technical mutation can be figured as telephone. The Old Testament unfolds a drama of listening and inscription of law; in contrast, the New Testament produces a kind of videodrome revision of some of the themes, topoi, and localities of its ancestral test. Now, I am referring to these texts not to admit to a conversion of any sort but merely to turn the dial and switch the ways our being has been modalized by a technology that works according to a different protocol of ethical attunement. If I refer us to the twin testaments, this in part is to explore the site of testimony that television has initiated.

It is no accident that television, in view of the dramas we have come to associate with the names of Lacan, Elvis, Heidegger, Anita Hill, Rodney King, Lee Harvey Oswald, Vietnam, Desert Storm, and so on, has become the locus of testimony, even if we are faced with false testimony or resolute noncoverage. In a moment I shall try to show why the Gulf War was presented to us as a discourse of effacement—why, at moments of referential need, the experience of the image is left behind. This has everything to do with the interruptive status of death and with the problem of thematization. In other words, there is a concurrent mark of an invisible channel in television that speaks to the problem of thematization, which makes the rhetoricity of the televisual image collapse into a blank stare.

At moments such as these—most manifestly, during the serrated nothing that was on during the Gulf War—television is not merely performing an allegory of the impossibility of reading: this would still be a thematizing activity (the problematic activity of thematizing will have been taken up again in the Rodney King case). During the Gulf War, television, as a production system of narrative, image, information flow, and so forth, took a major commercial break as it ran interference with its semantic and thematic dimensions. The interference television ran with itself, and continues to rerun on a secret track, points us to something like the essence of television.

The Rodney King event, which forced an image back on the screen, presented that which was unpresentable during the Gulf War—Rodney King, the black body under attack in a massive show of force, showed what would not be shown in its generalized form: the American police force attacking helpless brown bodies in Iraq. Now, the Rodney King trial was about force itself. Thematically, what was being measured, tested, and judged in the televised trial is the question of force as it eventuates in the form of excessive police force. And there's the catch, which we saw blown up and cut down in the Anita Hill case: force can never be perceived as such. This is a persistent question circulating in the more or less robust corpus of philosophical inquiry. The question becomes, how can philosophy talk about force?

The "theme" par excellence of the Rodney King show produced arguments concerning the regulation of force, its constitution and performance, which proved to be thematically disturbed and could be scaled only in terms of an ethics of dosage—escalation and deescalation of force. This demonstrated that we still do not know how to talk about force with an assured sense of its value or implications for judgment. While TV was under the covers, nomadic video captured images of brute force committed by the LAPD. Anyone who watched the trial knows that the referential stability of the images was blown out of the water: witnesses were reading blurs and blurring images. The status of the image as a semantic shooting range was severely undermined as TV conducted this interrogation of force. This interrogation, though, was forced upon TV—it involves an interrogation about its own textual performance in the production of force. What comes out provisionally, at least, is the fact that video, *nomadic* video, aleatory, unpredictable, vigilant, testimonial video emerges as the call of conscience of broadcast TV.

[289]

HEADLINE NEWS Expenditure, wasting time. Getting wasted. A bug or virus that started spreading after the war; the endless survival of what has not been fully understood. The circuits are loaded with the enigma of survival.

CHANNEL FOUR   What interests me are the two eyes of television. TV is always watching, always involved in and/or subjected to monitoring and surveillance when it has two eyes, one of which can be the eye donated by video. Not that video is the truth of television, or its essence: it is what watches television, it is the place of the testimonial that cannot speak with referential assurance but nevertheless asserts the truth of what it says. It becomes the call of conscience, which is to say, it responds in some crucial sense to the call of television.

Now, you have seen me play with the contrast and wave TV into the realm, or rather logic, of telephony. How can TV make a call, or more precisely, respond to a call? The way we call it is critical. (At the same time that TV is being watched by video and called to order, football has decided to dispense with instant replay. This decision, while made by team owners and not by media technologists, theorists, or media activists, nonetheless asserts that when it comes to calling it, TV withdraws its bid for claims made on behalf of referential stability.)

Let me try to unfold some of these points, and indicate where I think I'm going. On one level, television calls for a theory of distraction, which appears to be rooted in the trauma that it is always telling yet unable to fix. This suggests a complicated economy of visual playback and shock absorption, for trauma essentially involves an image without internalization. Cathy Caruth has argued that trauma does not "simply serve as a record of the past but precisely registers the force of an experience that is not yet fully owned.... [T]his paradoxical experience...both urgently demands historical awareness and yet denies our usual modes of access to it.... [W]hile the images of traumatic reenactment remain absolutely accurate and precise, they are largely inaccessible to conscious recall and control."[10]

[290]

HEADLINE NEWS   The relatedness of television to the end of history. Codes for such a reading were punched in when Benjamin and Freud discovered the experience of shock and the steps beyond the pleasure principle. This leads us to ask: What does it mean precisely for history to be the history of trauma?[11] It becomes necessary to translate the possibility of history by means of technology's flashback programming, or the prerecorded logic of TV.

Let me bring some of these strands into contact with one another. TV is irremissible: it is always on, even when it is off. Its voice of conscience is that internal alterity that runs interference with television in order to bring it closer to itself, but this closed-circuit surveillance can be experienced only in the mode of an estrangement. Television presents itself as being there only when it is other than itself—when it mimes police work or when, during the Gulf War, blanking out in a phobic response to the call of reference, it can become a radio. And yet, it's not that simple: in this case it *showed* itself not showing and *became* the closed, knotted eye of blindness. Within this act of showing itself not showing, posing itself as exposed, which is to say, showing that its rapport to the promise of reference is essentially one of phobia, it produces a dead gaze—what Maurice Blanchot would call "a gaze become the ghost of eternal vision." There is something in and from television that allows sight to be blinded into a neutral, directionless gleam. Yet, "blindness is vision still," "a vision which is no longer the possibility of seeing."[12]

In a sense, TV doubled for our blindness, even performed the rhetoric of blindness that guided the Gulf War. It entered us into the realm of the eternal diurnal, night vision, and twenty-four-hour operational engagement; its unseeing gaze was figured by the hypervision responsible for TV-guided missiles that exploded sight at the point of contact. Television showed precisely a *tele*-vision, a vision that is no longer the possibility of seeing. If it taught us anything, it was that what fascinates us also robs us of our power to give sense; drawing back from the world at the moment of contact, it draws us along, fascinated, blinded, exploded. Despite the propaganda contracts it had taken out, television produced a neutral gleam, one that tells us the relation between fascination and not seeing. If it showed anything, it showed a television without image, a site of trauma in which the experience of immediate proximity involved absolute distance. But it was through video, intervening as TV's call of conscience, as foreign body and parasitic inclusion in broadcast television, that a rhetoricity of televisual blindness emerged. Marking the incommensurate proximity of the same, testimonial video split television from its willed blindness and forced it to see what it would not show. Something was apprehended.

There was the undisclosed Gulf War. Elsewhere, I have tried to read this according to Lacanian protocols of mapping the maternal body— the imaginary mapping of the primordial aggression of the subject reproducing the cartography of the mother's internal empire: the origin of all aggression, according to Lacan and, before him, Melanie Klein, which explains in part why mothers became such prominent figures in this war that refused to figure itself materially. Mother ("mother of all battles," "mothers go to war," and so on) was not only put on the map by this war; she was the map. Can't get into her now, though it can be said that all forms of (paranoiac) aggression are perpetrated upon a displaced cartography of the maternal or, in the case of white on black, upon a body mangled by the rage through which the Other is marked with lacerations of the feminine.[13] As crude as it may seem to recognize this—and yet who doesn't know it?—Rodney King was put upon and sodomized ("Saddamized" in the metonymic citation) by brutal hets.

CHANNEL THREE The Gulf War was blanked out, put in a position of latency. As with all unsuccessful attempts at repression, however, the symptoms were bound to come rocketing out of a displaced area of the vast televisual corpus. The Rodney King event staged the survival of the Gulf War in its displaced form. Indeed, on his last day in office, Chief of Police Daryl Gates put the blame on the media—on TV—for precisely this type of displacement: "You made it into something bigger than the Gulf War." This holds for the beating; but when the troops were sent into L.A., the media reverted to their failure to show, recovering in essence their rerunning away, or the structural relation to the whited-out war.

What we call "Rodney King" has brought the question of force— officially, that of "excessive use of force"—to a hearing; it has placed police action on trial. Desert Storm was time and again cast as police action. Whether or not this represents a conceptually correct assessment in terms of strategy, buildup, tactical maneuvers, declarations of intent, and so on, it nevertheless remains true that at this instance the national unconscious inscribed the collapse of police action and military intervention. "Rodney King"—who as such never presented himself in the trial that refers us to him—names the hearings that never took place for war crimes committed in the Gulf War: condensed and displaced to the beating of Rodney King, the televised trial thematized unthematizable force fields of intensity while studying the problem of impact and the incitement to brutality with which TV has always been, in one form or another, associated. What is the relation between television and violence? Hasn't this been television's only question, once you get down to it? "That's not the way force is studied," retorted a use-of-force expert several weeks into the trial.

COMMERCIAL BREAK Bringing to the fore a study of force impact involving the difference between incapacitating strikes and pain-compliance impact devices (in particular the electric TASER gun), upper-body control holds, theories of escalation and deescalation of force, the trial induced by Rodney King showed how television, forced by video, was hearing out arguments organized around its own essence. This essence was shown to be linked to questions of trauma control and the administration of force. As the use-of-force expert tells it, on the first level the "empirical" force spectrum entails evaluations of verbal communication and effects of presence, while the second level involves responsiveness to pain-compliance impact devices, including upper-body control hold and the chokehold (whose routinely racist applications the police recognize and hence avoided using when beating Mr. King). Resolutely uninterrogated, however, the force spectrum fails to provide a reliable grid for evaluating force because it has not been tapped by theoretically sound means that would throw some light on what constitutes "effects of presence," "responsiveness," "communication," and so forth. I am not so foolish as to prescribe a mandatory reading list of Heidegger, Wittgenstein, Deleuze, Foucault, and Derrida in police training, though it would not hurt (their victims). Yet once the police started knowing or reading what they were doing, or whom they were representing and why, they would no longer be the police, the phantom-index to which Benjamin's argument pointed us. At the very least they would be, as readers, essentially detectives (and only essentially), those who, resisting group formation, sometimes have to turn in their badges or cross an ethico-legal line in order to investigate, piece together, read, and scour unconscious densities of meaning. It's not a pretty job, and it's generally

[293]

managed according to a different time clock than the one the police regularly punch into.

To the extent that pedagogy was blamed for the failure of the police to understand the scene of arrest or to control the usage of brute force—and that teachers were asked to speak about teaching—the trial focused on questions of how to read or, at least, how to produce effects of learning. While the introduction of the "force spectrum" was never in itself reflected upon or theorized, it nonetheless serves to circumscribe such levels of responsiveness as law enforcement and television, each in the idiom peculiar to it, attempt to elicit and regulate. Working over the arrested body, each inscribes and wastes it, making it do time—sometimes along the lines of teaching a lesson.

HEADLINE NEWS *If television and the police are experiencing some kind of shared predicament of being on call, it is necessary that we understand how these calls are circuited through the transmission systems of police force. The circuits are loaded: they range from the call of duty to which an officer is likely to respond with violence to highly suspect considerations of what constitutes responsiveness. (The defense's refusal to interpret Mr. King's calls as cries of pain was maintained throughout the trial.) Yet, the motif of the call ambiguates the scene of violence, for the possibility of appeal and the call of conscience are logged on to the same systems.*

Were it not that TV were itself trying to tell us something about the status of legal and social fictions, one would likely have to bring to bear a critique of violence in the manner of Benjamin. TV does not know what it knows: in the idiom of Heideggerian insight, TV cannot think the essence of TV, which it is, however, constantly marking and remarking. Its principal compulsion and major attraction comes to us as the relation to law. As that which is thematized compulsively, the relation to law is at once there and not there, canceling its program by producing it. (Hence the proliferation of police shows, from *Dragnet* to *Perry Mason, 911, Hard Copy, Top Cops, FBI, Law and Order*, and courtroom dramas; even westerns, with their lone law enforcer and inevitable sheriff, belong to this topos.) This relation to law, which television compulsively repeats as its theme, is simultaneously presented as the unthematizable par excellence—which is to say, this is a relation that cannot be presented as such but can only be appealed to or offered up as metonymic citation. Television is summoned before the law, but every attempt to produce the relation to law on a merely thematic level produces instead a narrative that is itself metonymic; the narrative is metonymic not because it is narrative, but because it depends on metonymic substitution from the start.[14] In other words, television cannot say the continuity of its relation to itself or its premier "object," which can be understood as force. This is why the Rodney King show is about television watching the law watching video; its call to order, a figure of order that tries to find the language by which to measure out an ethical dosage of force. At no point do television or policing delude themselves into assuming they can do without force, but they do not question this essential supplement. (Concerning supplement and dosage, television, as a drug, is also a tranquilizing force that regularly absorbs and administers hits of violence. After a hard day's work on *Psycho*, Alfred Hitchcock used to doze off in front of TV, claiming that TV, unlike film, was soporiferous.) Alternately stimulating and tranquilizing, ever anxiety-producing, television belongs to the domain of the internalized "posure" of drugs. This is why the Rodney King event has to start its narrative engine with a false start acknowledged by all: everyone involved in the chase had to start by assuming that they were pursuing a PCP "perp." Without this technology's relation to the asserted effects of drugs, hallucination, and supernatural force, there would be no act of television reading itself, no "self" pumped up on the base-free supposition of drugs.[15]

HEADLINE NEWS Implicit difference between tele-vision and video according to chapter seven of Being and Time*: Heidegger comments that the* scheinen *of semblance, in its various forms, "is founded on pure* phainesthai, *in the sense of phenomenal presencing. The Greek essence of* phainetai, *he notes, differs from that of the Latin* videtur, *even where mere semblance is denoted, for the Latin term is offered from the perspective of the observer, rather than from out of the unconstrained spontaneity of presencing.*[16]

HEADLINE NEWS The disruption to experience and comprehension that trauma involves.

What then, were the charges made in the Rodney King trial? The defense tried to show that, following the car chase, Mr. King took a step, which in fact was a charge—charging the LAPD. The counter-charges, made by the defense, pivot on the difference between a subject who is taking a step and one who is charging. If King was charging, then the force used to subdue him would have been justified. The distinction between taking a step and charging could not be determined with certainty by the footage provided in George Holiday's video. There seems to be an impasse, even though a phenomenal imaging of this scene exists. Repeated several hundred times in court, the frame-by-frame analyses explicitly raised questions about the relation between video recording and human memory. When witness David Love relied on his own memory of the beating, he felt the violence to have been entirely justified; however, when he was asked to interpret the video, he found the "same" scene to display an excessive and inexplicable use of force. Throughout the testimony it was asked of this witness to express "what does or does not the video say." The entire problematic of witnessing came into play. An assertive if provisional conclusion nominated the video as "the best witness"; but video, it was further argued, "doesn't tell the whole story" because it cannot reveal "state of mind." There were no strong readers around; at least none were called in. Superior Court Judge Weisberg ruled out the expert testimony even of psychologists (who should not perhaps be confused with strong readers).

Is there a whole story, a totality of a story that eludes the video scope but can be located elsewhere? Is there a state of mind, a clarity of intention, an interpretation with its totalizing impulses, upon which the LAPD can confidently count?[17] We know that a "strong reading" (one should measure how much force strong reading requires) of the tape would need first to account for these metaphysical ploys and rhetorical deceits, if only to discern the axiomatics upon which the constitution of force could be thought. What the video cannot in any case show, states the court, concerns an interiority, which it cannot inscribe; the video is pure surface without depth, running a mystifying release precisely because it fails to record inner perceptions. Unfueled by metaphysics, video is running on empty. Without access to interiority, the videotape deflects the scene from its locus in truth. That is why the court ruled in favor of human memory of violence, its flaws and gaps notwithstanding. Precisely where human memory of experience fails to achieve cognition—so the logic seems to go—it captures the "whole story." The court depended upon this evacuated site in order to retrieve a sense of the totality of the scene. This explains how the videotape's excess weighs in as deficiency in court. A mere machine, simply present while at the same time devoid of presence, it originates in a place without truth. As pure surface, the videotape effaces interiority as a condition of running. This is why the police give it chase, on and off the streets, in your face and behind your back.

Did Rodney King charge or step? The video that recorded this moment does not tell. The phenomenal instability of the image is staggering: there is no assured way to read the syntax of the move on a literal level. This step, which is out of line with all the certitudes we think we have about documenting the real, is in sync with Lacan's assertion that our encounter with the real is a missed appointment. In terms of the reading protocols that make up our legacy, the step that hesitates referentially between a step and a charge, tripping up the case as it does, is also a Freudian slip, a lapse and collapse in the grammar of conscious imaging. So the television watches as over this unreadable the video is on compulsive replay, tripping scene that it has witnessed. "*Gehen wir darum einen Schritt weiter*," writes Freud in *Beyond the Pleasure Principle*. Freud, like the video on compulsive instant replay, reruns this step that goes beyond the pleasure principle. Let us take another step.

HEADLINE NEWS Implications for memory: the difference between internalized memory and Memorex clips that run along the lines of flashback and other intrusive phenomena.

It is as if Freud is watching us watching this scene that returns incessantly. Running through psychoanalysis, vectorizing our thoughts toward "current events," in fact, we are looking at recurrent events whose eventuation cannot as such be easily located. Trauma reduces us to scanning external stimuli whose signals beam out a density of materials for historical reconstruction: "We describe as 'traumatic' any external stimuli which are powerful enough to break through the protective shield [layering].... The occurrence of a trauma externally induced is bound to create a major disturbance in the functioning of the organism's energy and to set in motion every possible defensive measure."[18]

The Rodney King trial in its particularity constitutes a moment when television reads itself and, staging itself reading itself, it is prompted by the interpretation of force set out in *Beyond the Pleasure Principle*, where the death drive kicks in by taking repetitive steps toward a beyond. It was Derrida who first noted, in *Post Card*, how Freud keeps on trying rhetorically to take another step in an attempt to get beyond a textual impasse. But going nowhere on a fast and invisible track keeps stepping up the momentum of external force, eventually achieving what Freud sees as the phenomenon of "breaking into the psychic layer that protects against excessive force." The dramatic incursion of excessive force peels down this protective layering, radically exposing the subject to the domain of the traumatic. In the realm of media technology, such a structure of protective layering has been historically provided by television which, up to a point, manages the scenography of external stimuli. The excess of the Rodney King intrusion upon broadcast television dramatized the rupturing of the protective film with which television habitually covers itself by showing and producing the traumatic scene of "excessive force." Broken in upon by testimonial video, television ceased to protect against that very thing it is intended to regulate. Formally on par with television, what Freud calls the domain of the traumatic is not as such a domain according to classical calculations of space and time but that which opens up a site of tremendous disturbances whose limits are difficult to discern. Television, the "domain" of the traumatic, while producing historical effects of reference, cannot be located in the world, but points instead to paradoxes of temporal complexities. We are on location, dislocated to the site of a provocation from the past that stammers over the *pas au-delà*—that Blanchotian space where the step can and cannot go beyond, restricted by a prohibitory injunction that points us backward as we attempt to trace the future of a step. The step beyond also involves the tripping that made it possible for the taped brutalization of Rodney King to blow out of teleproportion and into the streets.

HEADLINE NEWS *"The historical power of trauma is not just that the experience is repeated after its forgetting, but that it is only in and through its inherent forgetting that it is first experienced at all".*[19]

What urges us on and motivates linkage between Freud's text and Rodney King's text, and with that underpinning the war on drugs, and all the steps we have been impelled to take beyond the pleasure principle, involves the fact of fundamental disruption in traditional modes of consciousness and understanding, a disruption that occurs traumatically in the very experience of our history. This invasion of consciousness, a type of break in the possibilities traditionally allowed for experiencing experience, is what Benjamin called the *Chockerlebnis*—a jolt which occurs when an event is dissociated from the understanding that might attach itself to it; shock producing a split of memory from consciousness, often triggering technologically morphed mechanisms on the order of flashback or hallucination.

CHANNEL TWO The trial produced a number of maps, photographs, and flow charts of chronological time sequences; yet these common devices for capturing empirical parameters of events have failed to prove much of anything—except, possibly, that we are dealing with a type of experience that eludes temporal and spatial determinations altogether, something that can bust into a scene at any time, any place, miming the experience of the police. If the Rodney King beating figures the survival of the effaced Gulf War, then its principal "object" of projection would involve the phantom-text of a trauma. Precisely because the trauma is hidden from televised view—the Rodney King beating is a metonymy of a hidden atrocity, be this the unshown war or the atrocities to which African Americans are routinely subjected—it is accessible only by reading. The spectral trauma remains hidden even to the hidden camera that blindly captures it. Yet capturing the hidden trauma— not the suspect called "Rodney King" or even the police out of line— is how video has participated in focusing the disruption to experience and comprehension that TraumaTV involves. Under nocturnal cover, nomadic, guerilla video captures no more than a debilitating discrepancy (always screened by television) between experience and meaning that

Freud associates with trauma. This is why it could prove nothing but this discrepancy in a court of law.

As the trial tried to number the blows, count the strikes, and determine the velocity of force, all it can do is attempt to parry the shock that "in modernity dissociates once and for all the traditional cohesion of experience and cognition."[20] The repeat performance of a frame-by-frame blow shows how this text became nothing more than the compulsive unfolding of a blank citation. Video intervened as a distance that separates the witness's knowledge of the traumatic occurrence from the sheer repetitiveness that marks the experience of its telling.

CHANNEL ONE Is it accidental if one refers to the function of witness repeatedly by using the masculine pronoun? Or is it perhaps an "accident" of such magnitude that its enigmatic character has been somewhat effaced? Testimony, as Freud knew, reverts to the privilege of testicles, engendering truth within the seminal flow of testimonial utterance. Let's take this a few steps further. Standing as witness, in step with Freudian logic, and bearing testimony (*Zeug, zeugen*), swearing in the truth of one's testimony upon one's testicles, implies that the subject before the law comes under the threat of castration. The truth is related to this threat. Oedipus the video, lagging behind, limping out of step and out of line, plucking his eyes out when he sees the truth: this is the truth of video, the site of the neutral gleam that knows something that cannot be shown. When Freud traces testimony back to testicles he is also severing truth from any security net that might underlie cognition. Testimony, and that which it begets, is linked not so much to perception but to speculation. When Roman jurists swore upon their testicles they were swearing upon a truth that could never be known for sure, a truth whose resolution no amount of evidence could do more than

swear. Swearing, bearing witness, producing the testimonial—these constitute acts of language that, unfounded (that is, neither founded as in poetic speech nor grounded as in philosophical speech nor even secured by "ordinary language" usage), rely upon the vagaries of speculation, displacing the *testimony* to the fragility of the eyes: the two eyes of television and video, which are committed to the uncertain rigors of reading. Whether you are making sense or semen, you can never know for sure whether you are indeed the father of truth. Thus, in its essence and logic, testimonial is fragile, uncertain, performative, speculative. (In this regard, the one who is feminized, on the side of sense certainty, penetrated by force, the figure of excluded negativity, is bound to lose out to the symbolic inscription of the testimonial.) The legal mode of the trial "dramatizes...a contained and culturally channelled, institutionalized crisis of truth. The trial both derives from and proceeds by a crisis of evidence, which the verdict must resolve."[21] As a sentence, the verdict is a force of law performatively enacted as a defensive gesture for not knowing.

FACE THE NATION To this end, the Clarence Thomas hearings say more about that which cannot be presented, the relation between phallus and castration, the unrelenting crisis of evidence, and the nature of the testimonial as the drama by which the symbolicity of testicles come to be marked. These hearings bore witness to the powerful but empty phallus that could not be summoned to appear but around which the hearing was organized. This was not a negligible testimonial but one addressing itself to the essence of a supreme organ of state, namely, the Supreme Court of the United States. In this case, which tested the case of the case—the essence of testimony and the rectitude of justice—race, I daresay, initially disguised the sexual difference upon which legal testimony is erected and judgment based. It will not do to simplify the case by stating that Justice Thomas was a black man; identified as such, the African-American nominee carried with him phantasms of the *jouissance* of the other and effects of the phallus. In this regard, race aggravated the demand for presenting the phallus: but, like the phantom it is when presentation is at hand, the phallus was made to show up neither *in camera* nor on camera.

CHANNEL ZERO My contention is (and others have argued this according to different impulses and grammars) that television has always been related to the law, which it locates at the site of crucial trauma. When it is not performing metonymies of law, it still produces some cognition around its traumatic diffusions: thus, even the laugh track, programming the traumatic experience of laughter, can be understood to function as a shock absorber. It signals the obsessive distraction that links laughter to a concept of history within which Baudelaire, in "On the Essence of Laughter," located the loss of balance and, indeed, "mankind's universal fallen condition." With loss of balance and the condition of falling, we are back to that unreadable blur that is said to project the step—or the charge—taken by Mr. Rodney King on March 3, 1991.

"Trauma stops the chronological clock," writes Langer.[22] This stopwatch configures what makes television, despite the insistence of its "60 Minutes"–like ticking or the breathless schedule that it runs—freeze.

Still, television stops the chronological clock that it also parallels in a fugitive, clandestine way and according to two modes of temporal assignment. It stops time by interrupting its simulated chronology in the event of an "event" that is neither of time nor in time but, rather, something that depends upon repetition for its occurrence. The "event" usually enters television from a place of exteriority in which the witness is figured by an untrained video operator. And television stops the chronological clock by miming its regularity and predictability around the clock, running and rerunning the familiar foreignness of traumatic repetition. Indeed, one would be hard pressed to prove that the effectiveness of TV was not a symptom of the traumatic stress that TV also works to perpetuate. In an article on trauma, B. A. van der Kolk and Onno van der Hart describe trauma as if it were linked to the very functioning of the television apparatus, or, at least, as if the traumatized subject were caught in a perpetual state of internalized channel surfing: "He switches from one [existence] to the other without synchronization because he is reporting not on a sequence but a simultaneity.... A different state consists of a continuous switching from one internal world to another."[23] Television, a monument to that which cannot be stabilized, captures the disruption, seriature, the effraction of cognition, internal breaks—whether commercial or constitutive—and is scripted by the need to play out the difference between reference and phenomenality. This has everything to do with the essential character of traumatism as a nonsymbolizable wound that comes before any other effraction: *this* would be TV's guide—how to symbolize the wound that will not be shown.

Of the symptoms that television most indelibly remarks, one is the alternation marked between hypermnesia and amnesia. What is the relation between amnesia and the image? In films such as *Total Recall*, in order to discover the limits of any positable reality, acts of remembering are prompted by mnemonic devices along the lines of video implants. In fact, video has tracked considerable thematizations of internalized, commemorative memory. (*Erinnerung*, in Hegel's vocabulary, are nothing if not the literalization of *Gedächtnis*, an external memory prompter, a cue, or memo-padding.) While these video implants are often accompanied by nightmarish hues, they somehow remain external to the subject who needs these prompters to supplement an absence of memory. The image comes to infuse an amnesiac subject. "Total recall" is not the same as memory or recollection, and it is only total to the extent that it names the need for a prosthetic technology that would produce a memory track. In such films the video transport—these are always pointing to a modality of transport: constantly neurotransmitting highs, crashes, incessant repetition or fuzzing, they combine the idioms of drugs and electronics—the technochip induces some sort of trip, a condition of memory seen as lapses, stimulating the transmission of the slip. The video transport coexists with a condition of stated amnesia. It is to this amnesia, channel surfing through blank zones of trauma, that television, operating on screen memories and forgetting, secretly measures the force of an unbearable history.

The Haunted Screen

Kevin Robins

> There were 300 Americans killed in the Gulf War. During
> the same time, 300 Americans were murdered. What kind
> of country is this?
>
> —Jean-Luc Godard

What if we ran the tape back and replayed the Gulf War after *Psycho*?
What could this tell us about violence in Western culture? What would
it say about how we use the screen in our culture of violence? In this
essay, I am concerned with the interaction of technology, violence,
and fantasy.

Not so long ago—though it now seems an age—we were watching
the Gulf War being played out on our screens. We were spooked by
devilish images of Saddam Hussein, the "new Hitler." It was an epic
media event where the Good, armed with their high-tech weapons,
went in to "take apart" the Evil Empire. We did not see it, but we
came to know that more than 150,000 Iraqi soldiers and civilians were
slaughtered in that bloody kill. Then, just a few months later, we were
gripped by another kind of slaughter: the media turned to the phenomenon of serial killing in American cities. It is said that there are between
thirty and forty mass murderers at large in America, and their potential
victims were flocking into movie houses to be shocked and haunted by
knowing it. The monstrous image of *The Silence of the Lambs*'s Hannibal
Lecter was staring out at them.

We can use the serial-killing phenomenon to reframe the symbolic
value of the war. From the "deep black" of outer space, the penetrating
gaze of American spy satellites maintained a constant surveillance over
the Iraqi nation, continually identifying strategic targets. Stealth fighters
"lurked" in the Gulf skies, invisibly and undetectably lethal. And then
death came suddenly, out of the blue, through the "surgical strike" of a

cruise missile. These "smart" or "brilliant" weapons allowed the allied forces to "remove" chosen targets with precision and at will. And death came out of the black too. It came out of the night from killers who could see through darkness. One report describes how Apache helicopters, hovering fifty feet above the sand, "took out" a group of Iraqi soldiers: they were cut down by attackers they could not see as they ran with nowhere to hide and "did not know what the hell hit them."[1]

Disturbingly, something like this reality then seemed to come home to overshadow the populations of American cities. In the serial killer genre, it was the American citizen who came under surveillance, watched by the psychotic killer lurking stealthily in the urban night. The deadly, penetrating stare of Hannibal Lecter has become the motif of the genre. Lecter is characterized by his "brilliance," his "extraordinary brainpower": like the "smart" or "brilliant" weapon, he (serial killers are for the most part men), too, is programmed to strike precisely and at will. We cannot hide from the serial killer. He is in our midst, and he metes out death with apparent randomness: the victim is simply the wrong person in the wrong place at a particular time. We, too, have become a "bounty of targets."

Of course, there is a certain fortuitousness in this shift in agendas. The search for novelty and spectacle in the media is simply a random process. But, then again, perhaps there is more to it: in their juxtaposition, these two media phenomena reflect something important about violence in our culture and, particularly, about the screening (in both senses of that term) of violence. There is a compelling affinity: the sinister connotations of terms like "surgical strike" and "target-rich environment" are exacerbated in the context of urban mass murder in the United States. Phrases like "the silence of the lambs" cast a retrospective

shadow over President Bush's "black versus white, good versus evil" war in the Middle East. I think we should not sever the two media spectacles; together they tell us much about what film director Jonathan Demme describes as the condition of "moral disenfranchisement" that characterized "Bush's soulless America."

In an unusual essay on the Gulf War, Lloyd deMause, an American psychohistorian, describes this cultural condition as a "shared emotional disorder."[2] The war can then be seen as a kind of collective working-through of that disorder. America was experiencing feelings of guilt, depression, and sinfulness in the face of its own sense of chaos and impending dissolution; there was a need for expiation. In his efforts to illuminate the "inner life of America," deMause makes the apparently strange observation that it was child abuse that provided the "symbolic focus" for the crisis. Indeed, for some time before the conflict, he points out, the drama between the Terrifying Parent and the Hurt Child had been haunting the American psyche.

With the advent of war, the Terrifying Parent could be projected outward. It was Saddam Hussein who took on the role of child moles-ter. Western audiences were "sickened" by propaganda images of Saddam with his child hostages. Bush himself invoked the plight of the "innocent children." There were reports (which were subsequently shown to be propaganda fictions) that over three hundred Kuwaiti babies had died when Iraqi soldiers removed them from their incuba-tors. This was child abuse on a terrifying scale: Saddam was a violent monster and a merciless beast, the violator of every "civilized principle."

America identified with the Hurt Child, and Bush sought to take on the role of the Good Parent. This, as he made clear, was a war between good and evil, a war to protect and defend the "moral order" of the world. Saddam was seen as a contaminating and polluting force. As *Newsweek* magazine put it, "the chain had to be pulled, to flush Saddam away." In contrast, Bush's punitive violence could be seen as a cleansing and purifying force. What the war offered was the possibility of renewal and revitalization: America could rediscover its moral pur-pose and emotional wholeness.

Of course, there is a simplification in this account of the Gulf War as a kind of morality play; but there is also a persuasive truth in it. For a moment, a brief moment, this epic spectacle sustained a sense of national integrity and moral regeneration. The subsequent media event, the screening of serial killers, seemed immediately to question that self-confidence. These murderous films reintroduced an element of anxiety and dislocation into the national (and the Western) psyche: they brought the sense of disenfranchisement into the midst of the daily lives of Americans. They can be seen as retrospectively and insidiously poisoning the heroic image of the just war. Now, when we rewind the tape, the war looks different.

The Gulf War was to purify America by exorcising an evil that was projected as being outside, "in a desolate Middle Eastern desert." What became clear in the horror genre was the fear and anxiety that evil was really within. And strangely, yet not strangely, it was again the drama of the Terrifying Parent and the Hurt Child that was the symbolic focus of this anxiety. In *Twin Peaks*, Leland Palmer was the demonic murderer and rapist of his own daughter. In *The Silence of the Lambs*, the cannibalistic Lecter, himself a victim of child abuse, symbolically became the bad father of the detective, Clarice Starling. In John McNaughton's *Henry: Portrait of a Serial Killer*, the psychopathic Henry murdered the prostitute mother who abused him in childhood. If the American myth holds aloft the child as the symbol of innocence, there is something in that innocence that seems to arouse aggressive desires and drives in the parents who have charge of them. In America there was a wave of accusations about satanic practices, orgies, voyeurism, child rape, and murder sweeping the country.[3] Terrifying Parent and Hurt Child seemed caught in a vicious cycle of mutual destruction. But this time the destructive force could not be projected outward.

Creating a sense of impending dissolution and moral disintegration, it provoked feelings between anxiety and paranoia, between vigilance and panic. The films told us that this soulless America was Bush's moral order. The serial killer was the monstrous result of this moral world—and he could be anyone, the man next door, the next Henry or Jack or Norman that you had the misfortune to come across. He needed no reason or motive to kill you. Hannibal Lecter kills because that is his life, in a mockery of the idea of human motivation. The man who seems in control could suddenly explode into violent slaughter. The fictional Henry is driven by combined feelings of euphoria and dissociation: he kills because it makes him feel better. We saw this in the Gulf War, too: it was the monstrous result of this moral world. This time it was the Iraqis who had the misfortune to come across Henry, Jack, or Norman. Think of the Apache pilot returning from his mission:

"When we got back, I sat there on the wing, and I was laughing...
I lay there in bed and said, 'Okay, I'm tired, I've got to get to sleep.'
And then I'd think about sneaking up there and blowing this up,
blowing that up."

Perhaps we should consider what the difference is between this man,
who was capable of turning his hellish fire on men who were fleeing
"like ghostly sheep," and a killer like Hannibal Lecter, who devours his
victims. On the one hand, Iraq was a "turkey shoot"; on the other, it is
a question of cannibalism. The difference is that being Hannibal Lecter
in Iraq was all right; being Hannibal Lecter in America is not. Whereas
the Gulf War was projected as a simple confrontation between "us=
good" and "them=evil," in the serial-killer genre this simple moral
fiction is disturbed and confounded. In this genre, we recognize that
both the Hurt Child and the Terrifying Parent are aspects of ourselves.
We get to know what it feels like to be a victim. But it also brings
home to us how precarious is the line between us as victims and us as
potential aggressors. We too, under certain conditions, are capable of
unleashing a storm of violence. This genre forces us to confront moral
complexity and ambivalence. "Can you stand to say I'm evil?" asks
Hannibal Lecter. Lecter's question is a challenge to our moral condition: it exposes our emotional disorder.

But, how do we learn to live with this violence? To ask this question
is to consider the mechanisms through which we manage to screen ourselves from evil. In both of the examples I have been discussing, our
exposure to violence has been mediated through the screen. The screen
is a powerful metaphor for our times: it symbolizes how we exist in the
world, our contradictory condition of engagement and disengagement.
Increasingly, we confront moral issues through the screen, and the
screen confronts us with increasing numbers of moral dilemmas. At the
same time, however, it screens us from those dilemmas: it is through the
screen that we disavow or deny our human implication in moral realities.

I do not mean to say that this is intrinsically or necessarily the case.
Of course, we have all looked at images of suffering (in Vietnam, South
Africa, Ethiopia, Lebanon, Kurdistan—the list could be frighteningly
without end), and we have felt moral outrage. What I am saying is that

images present particular difficulties for our moral being. "To suffer is one thing," writes Susan Sontag, "another thing is living with the photographed images of suffering, which does not necessarily strengthen conscience and the ability to be compassionate."[4] Yet through the distancing force of images, frozen registrations of remote calamities, we have learned to manage our relationship with suffering. The photographic image at once exposes us to, and insulates us from, actual suffering; it does not, and cannot, in and of itself implicate us in the real and reciprocal relations necessary to sustain moral and compassionate existence. With video screens and electronic images, this moral chasm has been made wider. As we have become exposed to, and assaulted by, images of violence on a scale never before known, the affluent have also become more insulated from the realities of violence. It may no longer be a question of whether this strengthens conscience and compassion, but of whether it is actually undermining and eroding it.

If we are to come to terms with this moral condition, we must consider the nature of our engagement with screen culture. To do this, I want to focus on two episodes, one from each of the media events I have been discussing. The first is the report from the Gulf War, which I have already referred to, of Apache helicopter pilots video-recording their slaughter through cameras equipped with night-vision sights. After the carnage, the pilots returned to their base to watch the footage: "War-hardened soldiers hold their breath as Iraqi soldiers, as big as football players on the television screen, run with nowhere to hide." According to the report, the images showed one man who dropped, writhed on the ground, then struggled to his feet, until another burst of fire tore him apart. "A guy came up to me," says one participant, "and we were slapping each other on the back . . . and he said, 'By God, I thought we had shot into a damn farm. It looked like somebody opened the sheep pen.'"

The second episode I want to focus on is from *Henry: Portrait of a Serial Killer*. In this film, Henry and his friend Otis have stolen a video camera, and this is then used to video-record themselves torturing and slaughtering a suburban family. Afterwards, they sit back to watch a replay of their murderous and sadistic acts. "I want to see it again," says Otis, who likes the video so much that he eventually falls asleep in front of it.

The similarity of the two episodes is disturbing. What they dramatically emphasize is how peculiar the relationship is between the act of killing and the act of watching people being slaughtered. In both examples, the killers are, at the same time, the video makers. Acts of sadism are instantaneously transformed into acts of voyeurism. But what both of these episodes also show is that this dramatic transformation from sadism to voyeurism does not give rise to further moral dilemmas. As sadism burns into voyeurism it somehow neutralizes itself; in each case, it screens out the actual reality of the killing, and it distances the killers from moral engagement. How is this possible?

In the case of Henry and Otis, we know that this is because their behavior is psychotic: they are unable to differentiate between reality and fantasy. In their case, the reality and the image are simply substitutable. The question of moral responsibility is least problematic here, and we are concerned with how their voyeurism reflects the criminal pathology of the serial killer. But what of the Apache pilots? Surely they were not psychotic? At this point, perhaps, we should give up on the comparison and draw attention to the differences between the two episodes. There are, of course, clear and important differences, but I still want to push the similarities a little further. I think we can benefit from the opportunity that has fortuitously juxtaposed "normal" pilots and psychopathic killers. We can use it, perhaps, to consider the institution-alized and normalized defiance of reality that increasingly characterizes the military-information society.[5]

If the pilots were not psychopaths—and, for the most part, they surely weren't—we must consider how it was possible for them to screen out the reality of violence they had unleashed. How were they able to watch the videotapes and apparently dissociate themselves from the moral implications of their actions? And what is it that allows us to call their behavior normal? It is clearly the case that the pilots could find ways and means to achieve moral distance from the brutal and brutalizing activities in which they were engaged: they could see themselves as there to "punish evil." And through the defensive mechanisms of denial, disavowal, and repression, they could preserve and protect a certain sanity within an insanely violent environment.

One of the most powerful defensive strategies is the mechanism of *splitting*, which involves the division of the self and even the splitting of and disowning of a *part* of the self. In their book *The Genocidal Mentality*, Robert Jay Lifton and Eric Markusen describe how this mechanism worked in a profoundly violent environment, the Auschwitz death camp. Lifton and Markusen use the term "doubling" to describe how an element of the self can come to function autonomously and antithetically to the prior self: the individual involved in violence "perceived that the institution wanted him to bring forth a self that could adapt to killing without feeling himself a murderer. In that sense, doubling became not just an individual enterprise but a shared psychological process."[6] It is through this mechanism of splitting, the fragmenting of the self, that individuals can manage to coexist both as killers and as apparently normal people.

This process of splitting may be particularly important for understanding our implication in screen violence. We can describe the Apache pilots in terms of a splitting process that differentiates a spectator-self from an actor-self: in the context of the war, it became possible for them to feel that the spectator-self was more "real" than the self that was acting devastatingly in and on the real world. This was partly because the array of communications, control, and surveillance technologies in which they were immersed produced a kind of video-game scenario: it was a push-button, remote-control, screen-gazing war. The effect of these distancing technologies was to create a numbed experience of derealized and disembodied combat.

There was a sense of omnipotence and euphoria as the boundaries between reality and fantasy became disturbed. And, through this involvement in the screened war, the moral engagement of the actor-self in the reality of combat could be distanced and disavowed, for the more "real" spectator-self, subsequently replaying the video pictures of horror, served as yet another distancing device. The screen was the only contact point, the only channel for moral engagement, with the enemy other. At the same time, though, it amplified and legitimated the sense of omnipotence and power over that enemy. The screen was the only contact point, but what we must recognize is that in reality it was no contact point at all for moral engagement.

So, what of us? What of the rest of us screen-gazers, watchers, viewers, and voyeurs? Where are we in all of this? At this point, you may balk; you may feel that the comparisons are being overstated. Of course, I would have to agree in part: the behavior of television and movie audiences is different from that of the Apache pilots and of Henry and Otis. The individuals who make up these audiences are only spectators, not actors. It seems appropriate to describe their behavior as "normal," and it would seem strange to describe them as sadistic. The elements of difference are fairly clear. And yet, I would argue, the pathological still casts its shadow over the normal: the "ordinary" spectator of violence and suffering is not far removed from the extreme, the fantastic, aberrant, and frightening. In this respect, the screen has the potential to extend and amplify human awareness and sensibility. Of course, this can be liberating, but it can also be very problematic. The screen encourages a morbid voyeurism, a kind of bloodlust. Ignacio Ramonet has gone so far as to condemn the "necrophiliac perversion of television," the way that television takes nourishment from blood, violence, and death.[7]

The screen affords access to experiences beyond the ordinary. But experience and awareness for what, we might ask. What does it mean to be "fascinated" by a missile-eye perspective on death? What does it mean to become quickly "bored" by pictures of slaughter and suffering? What does it mean then to turn to horror movies to satisfy a need to be terrified? The spectator-self roves almost at random from one visual sensation to the next, a cruising voyeur.

The screen exposes the ordinary viewer to harsh realities, but it screens out the harshness of those realities. It has certain moral weightlessness: it grants sensation without demanding responsibility, and it involves us in a spectacle without engaging us in the complexity of its reality. This clearly satisfies certain needs or desires. Through its capacity to project frightening and threatening experiences, we can say that the screen provides a space in which to master anxiety. It allows us to rehearse our fantasies of omnipotence to overcome this anxiety.

[313]

In the serial-killer genre, this rehearsal may be about containing fears of our own destructiveness or about the impending dissolution of our civilized values. In the case of the Gulf War, the threat was (projected as being) from outside, and the screen was mobilized to construct a collective sense of omnipotence over an alien "monster." In their different ways, both of these media events of 1991 were about imagined threats to civilized norms and values, then about the imaginary exorcism of those threats. Screen omnipotence is about the drama of anxiety and containment. In the domain of the screen, it is possible to contain anxieties that cannot be confronted in their reality.

Moral identity and responsibility can only come from our recognition of, and engagement with, the refractoriness of the real world. It must necessarily involve us in processes of dialogue and negotiation, processes in which self and other are mutually transforming and transformed. The kind of screen event I have been describing is not characterized by such reciprocity: the screen bypasses the intractable nature of reality, and it seems to put us in control of the world. This is not to say that those who watch screens are not guided by moral concerns; rather, that the act of screen-gazing may make our moral behavior more difficult. The point is that the screen displaces (rather than supplements) reality: the very presence of the screen image testifies to the absence or remoteness of the screened reality. The screen is fundamentally inert: it does not involve us in the processes of dialogue and negotiation.

There are those (following Jean Baudrillard) who like to tell us that the screen has now eclipsed reality, that we are now living in a world of image, simulation, and spectacle. There is, indeed, something suggestive in this observation. But before we become too seduced by this postmodernist scenario, we should remember the 150,000 *real* men and women who were *really* slaughtered beyond the screening of the Gulf War. We should consider the implications of the fact that there is a symbiotic relationship between fictional serial killers and *real* ones who *really* slaughter.

It is not that we now live in the realm of the image; it is, rather, that there is, in our culture now, a kind of collective, social mechanism of splitting. The spectator-self is morally disengaged, floating about in an ocean of violent images. The actor-self is caught up in a reality whose violence is often morally overwhelming. How can we come to terms with this situation wherein the spectator and the actor seem to be going their separate ways?

Splitting
amnesia

The Gulf Massacre as Paranoid Rationality

Les Levidow

Technology as Ideology

Because this symposium concerns ideologies of technology, at the outset I would like to suggest a way of conceptualizing technology *as*
ideology. In a capitalist society the dominant ideology arises from the
commodity form. Its mysterious character was deciphered by Marx, of
course, who analyzed it in terms of fetishism and reification. Through
commodity exchange, he argued, the social characteristics of our labor
appear as "objective characteristics of the products of labor themselves,
as the socionatural properties of these things." We act as if social qualities and powers were attributes of things; in this way, we are controlled
by the instruments and products of our labor. Moreover, "To the producers, the social relations between their private labors *appear as what
they are*: i.e., they do not appear as direct social relations between persons in their work, but rather as objective relations between persons and
social relations between objects."[1] Thus, the term "ideology" refers to
our lived experience, our reified reality, not a misperception of the reality.

Herbert Marcuse extended this concept of ideology to technological
rationality and attacked Max Weber for idealizing instrumental rationality as a value-free, calculable efficiency; such an ideology legitimates a
specialized scientific administration that dominates nature and humanity
by reducing relations to quantities of things. Here Marcuse identified a
"reification of reason, reification *as* reason": everything is reduced
to the administration of things.[2] This formal rationality, moreover,
abstracts properties from their context, giving us reified *parts* of objects.[3]
Decontextualizing practices of formal abstraction transform objects of
administration—notably people—into mere means for whatever purposes may be imposed.[4] The Nazi Holocaust epitomizes the rational,
efficient processing of people as things.

Today, similar processes are facilitated through computerization, by which we reduce objects to digitized things. By attributing social powers to computers, and by fetishizing electronic "information" for its precision and omniscience, we come to judge ourselves against the mechanical or cybernetic qualities with which we have imbued computers. An operator *does* behave as a virtual cyborg in a real-time, man-machine interface, regardless of whether he or she structures military weapons or children's games and educational programs.[5] Some innovators of interactive simulation have alternated between designing military and entertainment versions of similar technology.[6]

Through such computer simulation, enemy threats—real or imaginary, human or machine—become precise grid locations, "targeting information," abstracted from their human context. It permits us to experience the world through a fantasy of omnipotence, of total control over things: Nintendo games may privatize such fantasies, but events such as the Gulf War publicize them, in the dual sense of validating and universalizing them. Fantasized omnipotence, however, also requires containing anxieties of impotence and of vulnerability to unseen threats.

It has been argued that as computer-imaging techniques become more sophisticated, their simulation capacity could devalue the objective authority of images generally. In terms of ideology, however, we should concern ourselves less with an image's "truth" value than with the power relations such images reify. Their political effect, in any case, depends less on their technological origin than on an entire cultural rhetoric.[7] The Gulf War has provided us with a rich example. Computer-mediated images helped achieve a deadly persuasion, such that a series of massacres was portrayed as heroic combat. How was their cultural rhetoric constructed, long before the United States attacked Iraq?

[318]

Stopping the New Hitler

The West effectively mobilized support for this "just war" by casting it as a rerun of World War II—or by casting it as a way of *preventing* another World War II, to *prevent* a new Hitler. We experienced Iraq through the metaphor of the state as a person, as if in a fairy tale of rescuing a victim from an irrational monster.[8] In the United States, the mission took on triumphant tones from the very start: displaying American flags and yellow ribbons, Americans itched to "kick ass," to overcome the "Vietnam syndrome," while preventing a Third World country from rising above its station.

Opinion polls showed an even higher proportion of popular support in Britain than in the United States, but with an ambivalent content; the support seemed more a resigned acceptance of an unavoidable task. It is notable that in late 1990, opinion polls indicated a "gender gap," with relatively more men than women supporting a Western military attack.[9] The gap narrowed when support overall increased after the United States attacked in mid January 1991. Rather than resort to essentialist explanations, such as male aggression, this essay will cite cultural clues that reveal subconscious fantasy being mobilized in the demonization of Saddam Hussein. The mass media was particularly significant for the obvious reason that most people experienced the war mainly through its words and images.

Saddamized Victims

After it was claimed that Iraq had "raped" Kuwait, that key verb entered common parlance. Although there were reports of Iraqi soldiers actually raping residents of Kuwait, the term referred more symbolically to a country forcing a weaker one into submission. This metaphor displaced undeniable facts of preinvasion Kuwaiti society, for example, that some Asian residents had served as the sex slaves of their employers and that immigrant workers had been denied basic civil rights. Rather than identify with these longstanding victims, we were encouraged to identify with an abstraction called "the Kuwaiti people." Their "rape" symbolized the threatened buggery of the West itself. After the war began, an oil troubleshooter from Britain said, "I wouldn't want them to bomb me—to send an Exocet up my backside."[10]

Iraq's relatively advanced military technology, too, was experienced as a threat to the West's sense of cultural superiority, particularly to its masculine identity. The implied threat to masculinity became more explicit when captured allied airmen denounced the West on television. According to one tabloid newspaper, they were "forced to take so many drugs that they are 10 percent of the men they really are."[11] Saddam Hussein himself came to personify the symbolic threat of buggery: in his notorious appearance with Western hostages, when he stroked the hair of a young boy, he became the living image of the predatory pederast.[12] The subconscious fear of buggery served well to mobilize Western males. More generally, he evoked a widespread anxiety of moral violation, even chaos. As if his real barbarity were not sufficient, newspapers and television circulated a story of soldiers evicting babies from incubators, since discredited as fabrications, and far less credible tales of Iraqi soldiers as "vampires," draining the blood of civilians.

With such imagery, the Hitler analogy portrayed Nazism as irrational and sadistic; it presented Iraq as a specter of "medieval" violence, almost naturally derived from irrationality. As Major General Julian Thompson speculated on Saddam's military strategy, "He is unlikely to approach the problem in what some in the West would regard as a rational way."[13] Conversely, of course, the West threatened and employed a morally based violence, characterized by its precision and rationality.

Yet as the West's war aims transparently moved beyond the official mission to free Kuwait, the ensuing destruction resonated with popular wishes to castrate the bugger, to annihilate the source of primitive anxiety. Castration metaphors arose in military language, which spread, of course, into common parlance. Journalists and politicians spoke of the need to "neuter" the Iraqi armed forces. Military strategists emphasized the necessity of winning "air supremacy," not merely superiority. Said Major General Thompson, "It is vitally important that the Iraqi air force is rendered impotent before ground battle starts."[14] Announcing his plans to attack the Iraqi army, General Colin Powell said, "First we are going to cut it off, then we are going to kill it."[15] U.S. Colonel Alton Whitley boasted, "You pick precisely which target you want— the men's room or the ladies' room."[16]

Reified Killing

The U.S. government explicitly targeted the enemy infrastructure, as if it were separable from Iraqi daily survival.[17] Bombed facilities were rhetorically personalized as Saddam's military machine. Reports routinely celebrated the successful targeting of "his" communications, bridges, power stations, and so on. As Robert Stamm has pointed out, "the media turned Iraq into one vast faceless extension of [its] demonized leader."[18] Paradoxically, the video imagery that brought us a visual proximity between weapon and target permitted greater psychological distance. This remote-intimate viewing extended the moral detachment that typifies earlier military technologies.[19] The "smart" bomb footage homing in on a target allowed war to *appear* as it is, conducted as social relations among things.

The language adopted by the media—again developed from military euphemism—attributed human qualities to inanimate objects while denying them to the war's victims. "Smart" bombs "killed" Iraqi equipment and Arab "cockroaches." The video-game audience morally detached themselves from the human consequences of surgical strikes against the evil, nonhuman forces personified by Saddam Hussein.[20]

As Iraqi conscripts were massacred, video-game images portrayed the carnage as a precise medical operation. "The operation was surgical, with no loss of American life," reported Julie Flint.[21] How did war acquire this moral medical language? Historically, modern medicine itself has been constructed from military metaphors: medicine devises surveillance and targeting techniques to attack the command and control system of disease, thus protecting the body from alien invasion.[22] Medicine's military metaphors came full circle in the Gulf War.

Instrumental rationality, too, came full circle. Through the capitalist epoch, a cult of efficiency has increasingly displaced issues of morality. Conversely, in the Gulf a rhetorical precision and efficiency provided moral legitimation for the U.S. attack. Our weapons were exterminating angels.[23]

What did the precision mean in practice? The United States was supposedly targeting the fearsome Republican Guard, which had spearheaded the "rape" of Kuwait, and which epitomized the barbarism of "the new Hitler." In reality, those troops had quietly left the front line before the mid-January deadline set by the United Nations Security Council. The blunt carpet-bombing proceeded to kill mainly Shia and Kurdish conscripts; in fact, Saddam had already "cut off" supplies to those on the killing fields, even before the U.S.-led attack began. Paradoxically, the Republican Guard was largely spared, so that they could prevent Iraq's democratic forces from overthrowing the Ba'ath regime. If there was any surgical precision in the U.S. strategy, then it politically targeted the same forces that Saddam had been suppressing before his invasion of Kuwait. Yet the "Nintendo war" images portrayed the massacre as honorable combat, as proxy attacks on the Ba'ath regime.

Video Game as Paranoiac Environment

The Gulf War was constructed as "total television," an entertainment form that merged military and media planning.[24] This infotainment engages its audience by evoking a familiarity with video games. The real-time TV images inverted an emotional resonance, which Gillian Skirrow had described a few years earlier: "The actual performance required of us in the video game is like being permanently connected to broadcast television's exciting live event."[25]

Analyzing boys' neurotic compulsion in video games, Skirrow conceptualized a "paranoiac environment" as dissolving any distinction between the doer and viewer. Driven by the structure of the video game, the player is constantly defending himself, or the entire universe, from destructive forces. The play becomes a compulsive repetition of a life-and-death performance.[26]

[322]

Skirrow drew upon Melanie Klein's theories to explain how such games reactivate the unconscious. Boys attempt to master anxiety over threats to the insides of their own bodies through an omnipotent penis-as-magic-wand. As a symbolic organ of perception, this discovers what destruction has been done to the insides of one's mother's body and what perils it faces, the danger turns from an internal into an external one. Consequently, "his fear of not being able to put things right again arouses his still deeper fear of being exposed to the revenge of the objects which, in his phantasy, he has killed and which keep on coming back again."[27]

His anxiety is never mastered by that vicariously dangerous play; he engages in a characteristic repetition, as expressed in video-game "addiction." The fantasy of penile "perception" suggests, then, a resonance with the omniscience, omnipotence, and paranoia associated with the remote video-computer targeting. Beyond these phallic-based fears, specific to boys, Klein describes more primitive anxieties, including the splitting of internal objects into good and bad parts, the destruction of the bad, the anticipation of retaliation, and the fear of annihilation of the self.

Aggressive feelings cannot overcome such fear as dreadful elements take the form of unnameable entities inside the self. Thus, apparently rational justifications for war always "conceal the underlying, unconscious part of the institution that seeks out or even invents the enemy... War is organized as a double security operation in which an implicit protective system underlies an explicit one. The hidden part of the war institution...acts to convert schizoid or psychotic anxieties, generated by illusory internal dangers, into perceptions of apparently real external dangers."[28]

Such psychodynamics arose in images of Saddam as the personification of sadistic, unpredictable, limitless violence. By exaggerating claims about Iraq's nuclear weapons development and speculating on its possession of chemical weapons, the mass media portrayed Iraq as a regional aggressor, posing a global threat of annihilation; a nameless dread was evoked by a "madman" who transgressed the combined rules of morality and rationality. Conflating internal and external threats, our own rationality became inseparable from a paranoid projection.

By considering these unconscious processes, we can begin to understand how video images in the Gulf War did far more than sanitize death; for, as Stamm has noted, "Telespectators were made to see from the bomber's perspective."[29] We became involved through TV images as vicarious participants in the destruction of perceived threats to our bodily integrity, our physical existence, and our social order. Indeed, we could feel a pleasurable identification with high-tech violence against an uncivilized enemy.[30] These perceived threats infantilized us, leading us to welcome a strong savior who was apparently wielding a civilized violence on behalf of international law.

The war imagery both evoked and contained primitive anxiety through forms of technological rationality, permitting us to experience the West's barbaric crusade as if it were asserting civilized values over barbarism, asserting rationality against irrationality.

Electronic Orientalism

How was the cultural rhetoric specific to an Arab target? Traditional Orientalist stereotypes were in themselves inadequate for dehumanizing Iraq, which was among the most secular, highly educated, and techno-logically advanced of the Arab countries. Previously, the sexual metaphors that pervaded Orientalist discourse portrayed enlightened Westerners penetrating the resistant virgin, seducing the Arab into modernization. Short of such uplift, Arabs remained essentially biologi-cal beings possessing an undifferentiated sexual drive. With the rise of Arab nationalism, revolutionary activity was portrayed as a bad sexuality, as an incoherent sexual activity rather than as politics.[31] During the Gulf War, Saddam was demonized as a unique threat of irrational violence, of undifferentiated or even sadistic urges, symbolized by sodomy; he represented a perverse, and well-armed, sexuality.

In addition, Iraq's cautious military approach was portrayed along the lines of the backward Arab as coy virgin. When Saddam decided to avoid a direct military confrontation with the U.S. coalition's air force, he was "hunkering down," almost cheating the surveillance systems of the West's enlightened, rational game plan: "Saddam's armies last week seemed to be enacting a travesty of the Arab motif of veiling and con-cealment...[Saddam] makes a fairly gaudy display of mystique," wrote *Time* magazine.[32]

These racist images clearly emerged after the U.S. bombed civilians in the Amariya air-raid shelter. In this case, TV images unusually pic-tured hundreds of shrouded corpses. In response, the U.S. authorities insisted that they had recorded a precise hit on a "positively identified military target"; they even blamed Saddam for putting civilians in a bunker.[33] The United States could only invoke its surgical precision as moral legitimation, even though it was the precise targeting that allowed the missile to enter the ventilation shaft and incinerate all the people inside the shelter.

Yet the U.S. response had an underlying logic: any attempt to evade penetration by the West's high-tech panopticon simply confirmed the guilt and irrationality of the devious Arab enemy. Any optical evasion became an omnipresent, unseen threat of the unknown that must be exterminated. This paranoid logic complements the U.S. tendency to abandon the Cold War rationales for its electronic surveillance and weaponry, now redesigned explicitly for attacking the Third World.[34]

According to Edward Said, "It is as if an almost metaphysical need to rout Iraq has sprung forward…because a small non-white country has disturbed or rankled a suddenly energized super-nation imbued with a fervor that can only be satisfied with compliance or subservience from 'sheiks,' dictators, and camel jockeys."[35] That "need" can be understood as containing an internal threat of chaos and annihilation— a threat equated with any Third World independence, and containable only by imposing total submission, even extermination. Although that "need" did not cause the war alone, racist anxieties were mobilized to accept or even demand it.

High—Tech Barbarism

In an article written just after the Kuwait invasion, ex-CIA agent Philip Agee compared the Gulf crisis to the Korean one in 1950. Both of them were border crises just waiting for escalation; both U.S. responses were justified as a defense of "our way of life."[36] As in Korea, the United States heightened tensions in the Gulf long before the Iraqi invasion, thus intensifying disorder to justify global intervention and continued arms expenditure.

In the late 1980s, when the Soviet Communist devil was disappearing, the U.S. military justified buying its latest toys by claiming to protect us from uncertainty. Mere "uncertainty" may seem a poor substitute for a pervasive threat of international Communism; however, through an extensive surveillance technology and video images, any country can become a simulated enemy, digitally reduced to a "target-rich environment."[37] Thus, we dehumanize people as mere things; a language of hygienic annihilation allows us to distance ourselves morally from the human consequences.

By comparing Saddam to Hitler, our society denies the veritable Western sources of Nazism, particularly the West's Faustian worship of technique. Instead we project Europe's murderous racist legacy onto the Third World.[38] While claiming to protect the world from new Hitlers, we extend the bureaucratic-technological rationality of the Nazi exterminations.[39]

As with earlier efficient massacres, our efforts to distance ourselves psychologically from the victims are belied by the sexual metaphors, expressing primitive anxieties with which we do so. By symbolically castrating the bugger, we contain the feared submission to a perverse sexuality, and even a feared collapse into psychic disorder; we project disavowed, destructive parts of ourselves onto thinglike enemies.

Although such paranoid fantasies alone do not explain war, their mobilization does indicate how readily we become infantilized, yearn for a savior, and treat our aggression as a self-defense. As spectators we can experience high-tech barbarism as if it were honorable combat, and even imagine ourselves guiding the weapons with surgically clean hands. If we are to resist such complicity, we will need to confront the paranoid rationality that evokes and contains our fears.

During the Gulf massacre, a paranoia structured the instrumental rationality of military action. A racist, reifying language disguised its nominal goal of saving humanity from barbarism. Although support for the war expressed desires for social collectivity, it remained psychopathological by idealizing our own murderous aggression as angelic. Thus it is misleading to suggest that public support represented "the desire for a communitarian solidarity with other people."[40] Indeed, war supporters often rejected information about the war's victims.

The problem here is less about people accepting the official "truth" of propaganda images than about their seeking refuge from uncertainty.

The danger is that "people will choose fantasy, and fantasy identification with power, over a threatening or intolerably dislocating social reality."[41] Such fantasy now becomes particularly appealing: we are told that the West won the Cold War but we wonder what we have gained and fear that we have much to lose in the New World Disorder.

Beyond Paranoid Rationality

The Gulf War illustrates the role of high-tech systems in mass psychopathology. A paranoid rationality expressed in terms of the machinelike self combines an omnipotent fantasy of self-control with fear and aggression directed against the emotional and bodily limitations of mere mortals.[42] As this episode illustrated, computer systems can seduce us into a participatory paranoia, turn our selves into social and emotional cripples, and extend a commodity-type reification far beyond market relations, to our very sense of who we are. This dynamic refutes the naive hopes of those who have idealized electronic information—as an instrument of participatory democracy, as a social prosthesis, or even as resistance to the commodity form.[43]

Electronic information, rather, operates as ideology, contributing to and accelerating the development of a paranoiac environment; it structures objective relations between people and social relations between objects. If we are to subvert such reification of our collective social labor, then we will need somehow to dereify technology, to appropriate its potential for mediating social relations between people. In doing so, we will need to find playful ways of addressing our primitive anxieties, vulnerable to being manipulated within technological systems.

The New Smartness

Andrew Ross

This is how it used to be in a world where people were always being warned not to be too smart for their own good: smartypants, smartass, smart alecks, the smart set, street smarts, smart dressers, supersmart, smart kids, get smart, man smart, woman smarter, smart cookies don't crumble, if you're so smart, then…

This is the way things are now when people are reminded that they are going to be outsmarted by virtually everything they come in contact with: smart buildings, smart streets, smart cards, smart drugs, smart fluids, smart food, smart bars, smart kitchens, smart docks, smart tunnels, smart highways, smart money, smart sensors, smartware, smart weapons, smart cars, smart windows, smart yellow pages (everything but smart presidents).

If you consult the *Oxford English Dictionary* you will see that the term "smart" has been through quite a few semantic changes in its time, but most would agree that the word has only come into its own in recent years, and is currently the busiest of buzzwords in the cutting-edge vernacular that embraces corporate think tanks, government planning agencies, New Age manifestos, and promoters of new restaurants and dance clubs. The term now irradiates the object world, whereas once it was an exclusively subjective quality. Given the recent attribution of smartness to things, it is tempting to consider whether this semantic shift does indeed register a significant shift in the ideology of technology. In the pages that follow, I will discuss this proposition, and consider how our own attitudes toward the new smartness may have a bearing upon versions of the New World Order that are currently being scripted and produced on the global media stage.

It used to be that only humans were smart—and some animals too: smart dogs, smart beavers, smart foxes (dolphins, of course, are not smart; they are "intelligent"). But it wasn't always a good thing for humans to be smart, especially if you were of the wrong gender or the wrong race, in which case you were automatically considered too smart for your own good. The attribution of smartness was as much a way of regulating the social potential of brainpower—a way of keeping people in their place—as it was a way of recruiting only the right sort of native intelligence for God and country and Dow Jones.

It also reflected tellingly upon the tradition of anti-intellectualism that is often thought to be endemic to North American culture. After all, anti-intellectualism draws its validity as much from below, from longstanding populist fears about the authority of a highly cultured elite, as it derives its power from above, from an antagonism of the corporate class to the traditional challenge of the left-liberal intelligentsia. In addition, anti-intellectualism is often employed in the same way that the charge of anti-Americanism, or even that of anti-Semitism, is used to exclude dissenting views as if they were irreducibly prejudiced rather than grounded in opinion. The resentment of intellectualism is not a species of know-nothing intolerance endemic to the U.S. populace; on the one hand, it is a cogent popular response to a material history of elitist privilege; on the other, it is a more or less structural result of conflict between factions of the ruling elites.

Smart vs. Intelligent

Has anything in this anti-intellectual equation changed with the advent of automated intelligence in the form of smart objects? On the face of it, populist resentment has no more dissipated than has the fierce passion of the governing elite to rein in, silence, or contain the intelligentsia. What has happened, nonetheless, is that the human-made object world has increasingly become an alternative home of intelligence, though hardly a substitute, except in the Wayne's World of corporate advertising utopia. Now, there may be nothing new in this: after all, you may say, smart machinery has merely become part of the new high culture, part of the new surveillance culture, part of the new regulatory system that keeps people in their places and regiments their work hours. For instance, there is every reason for workers to be as suspicious of automated intelligence, timing their every trip to the toilet, as they were of the human folks who used to make the decisions and give the orders.

So, too, you might say, is smart machinery simply a way of appropriating the function of the knowledge class, in much the same way that industrial technology (IT) once appropriated the know-how of artisans and semiskilled laborers. Why bother with these pesky intellectuals, when processing machines can deliver their mental product much more efficiently and without so much social spillover? The history of the development of information-processing systems is intimate with the rise of the professional-managerial class in this century. Among other things, both information technology and the professional-managerial class are institutional formations for stabilizing and controlling the predictive and planning functions of Gramsci's broad category of "intellectuals" —all of those who use words, statistics, or knowledge skills to organize and shape human activity. The result is a broadly entrenched technocracy, which commands a vast resource base of human and electronic facilities, and which enjoys the support of a pervasive ideology of problem-solving expertise and rule-driven competence.

The new smartness is native to the technocrat's house style. Smart, after all, is not the same as intelligent, let alone intellectual. Smartness is intelligence that is cost-efficient, planner-responsible, user-friendly, and unerringly obedient to its programmer's designs. None of the qualities, in other words, which we associate with free-thinking intellectuals. This is why these tedious debates among cognitive psychology types about whether computers really could play chess as well (or as creatively) as humans were all so much hot air. It was assumed that since chess was a game for intellectuals, computers that could master the game might be designated as worthy surrogates of human intellect. On the contrary, it's clear that what was really going on was the training of artificial intelligence (AI) for wargames, of which chess is simply a stylized civilian version. What did you think those Cold War showdowns between U.S. and Soviet grandmasters were all about anyway? Certainly not chess. The whole point of AI is not to stimulate human intelligence, never will be: the point is to create a more obedient, scruple-free, nonneurotic, and anatomically correct form of intelligence.

In brief, then, it is fair to say that the grand old tradition of anti-intellectualism, as I have described it here, has scarcely faltered in the new age of smartness. Indeed, AI and IT have been instrumental in appropriating the knowledge of the knowledge-class in the event of the material absence, replacement, or withering away of that class. This strategy—to replace what Alvin Toffler called the "cognitariat"—is a logical outcome of a postindustrial order that is driven and maintained by controlling or rationalizing the production, distribution, and consumption of an information commodity. Accordingly, what we have seen is a concomitant rise in the level of technophobia on the part of intellectuals and experts who rightfully resent seeing their accumulated cultural capital being transferred to a "knowledge base" in preparation for a more fully administered society.

As for the other side of the anti-intellectualist equation, there has also been a marked increase in their patronizing of the less well educated in the classic style of blaming the victim. The dominant prescription here seems to be this: as machines get smarter, people get dumber. This tendency, compounded with the latest round of national hysteria about educational standards relative to the Japanese and the Germans, the new trade-war rivals, threatens to revive the schooling panic that followed the launch of Sputnik in 1957. It doesn't take much to notice that standards, almost by definition, are always falling. We seldom see anyone, least of all in the field of education, get up and announce, with gusto, "standards are rising!" This is not to gloss over the postindustrial creation of a knowledge hierarchy, structured by uneven access to information. On the contrary, it is to recognize that what is perceived as "dumbness" is an intentional effect of this hierarchy, not its cause.

If we are going to talk about dumbness, then let's take our cue from the machines themselves, for if indeed things are getting smarter, then it's also true that they look a lot dumber. All smart machines these days come in very dumb boxes, uncommunicative containers that don't seem to say anything at all about their contents. The golden age of industrial design, at least from a fine-arts perspective, seems to be long gone. As opposed to those days when the design of an object was a stylish commentary upon its function, most information technology today is —what? A flat circuit board inside a generic box with no ostensibly expressive form. This is what happens when technologies go mental.

As boxes go, these technologies no longer even need to camouflage their purpose, like, say, the Tardis police box in the Dr. Who television series, or to advertise the mystique of their inaccessibility, like the megalith in the film *2001: A Space Odyssey*. Today's dumb box is basically a control board for regulating processes that cannot be physically represented as actions in geographical space or linear time. The planned obsolescence of these machines is no longer subject to external codes of style and taste—what could be called the "Detroit Principle." Obsolescence today is governed by the generational law of intelligence, embodied in progressively smarter chips. Consequently, machine aesthetics are still in some way determined by the expendability of objects, living, as we still do, in a throwaway economy. The difference is that computational intelligence is not all inert, so information and memory can be retrieved and transferred and reconstructed in ways that have challenged our deep-seated belief in the eventual death of machines. The dumb box is not then a coy masquerade of inertia, it is a haughty show of neutrality in an object world whose physical laws barely impinge any longer on the transcendent processes concealed within the box. In this respect, the dumbness of the box is expressive, not of the contents, but of the intentions of their creators, who arrogantly view their creations as some new and superior life form.

If the human-made object world—what is often called "second nature"—is increasingly smart and, inversely, less inert, then what does this say about our attitudes toward the first world of nature? For all sorts of good reasons, it's been quite a while since we thought of the natural world in the terms bequeathed to us by the Enlightenment legacy of scientific materialism—as an inert mechanism whose workings would be mastered by humans for their use alone. As most readers of this essay will probably recognize, humanism is a rather corrupt idea at this point in time, and you don't have to be an active eco-warrior in the ranks of Earth First! to understand why.

Alternative Responses

So what are the alternative responses today to the corrupt traditions of humanist thought? I can think of at least three—radical humanist, radical technologist, and radical ecologist—and I will devote the rest of the article to briefly outlining their claims.

Radical Humanism

The first emanates from the human-potential movements that are often associated with the New Age but which also can be said to embrace the more creative or maverick pioneers in technologies like virtual reality. These movements are devoted to reconstructing, enhancing, boosting, and upgrading humanism, employing technologies to build a new species, the New Prometheans unchained, that will be free—free at last—from the fetters of Nature. Whether this involves tapping the potential of unused DNA in the human biocomputer or maximizing the unused neurons that make up an estimated ninety percent of brain capacity, the aim is to reverse entropy and to usher in an evolutionary quantum leap for the species.

Nowhere are these aims more articulate than in *Mondo 2000* circles, where Timothy Leary's lifelong philosophy of mind expansion—in spite of everything, one of the most consistent philosophies of liberation in our times—has found an enthusiastic reception and a creative home. This crusade to make humans smarter has its own designer drug counterculture—the various cognitive enhancers, empathogens, entactogens, psychotropic drinks, and psychoactive foods. It has its own holistic rules of thumb, creatively cobbled together from the most ancient religions and the most advanced developments in neuroscience. And it provides a good deal of energetic lobbying and ethical support for the latest advances in biotechnology and bioengineering. Atom-stacking, enzyme design, biological self-replication, genetic assembly, protein computing, nanomachinery, and endorphin control—these are the principles of molecular technology that will lead us into the Age of Smartness, whether on Earth or on Spaceship Earth.

Despite its boundless optimism, there are reasons to be skeptical about the brave new world of this radical humanism, reasons to believe that this is one form of libertarianism that will only free a minority of humans—those at the cutting edge of the new smart technologies—and that its willful antagonism toward limits pays too little heed to ecological concerns. If the right to be intelligent really were to be recognized as a human right, say, by some enlightened bloc at the U.N., then it is unlikely that it would be posed as a corrective to the failure of individuals to tap their full biopotential. It would more likely be posed as a corrective to the world power system that links structural undereducation, illiteracy, and poverty with the structural overdevelopment of First World elites. On the other hand, we should, I think, hesitate before such a literal interpretation of what is essentially a utopian injunction on the part of an experimental counterculture. The function of countercultural claims is not to provide social blueprints but to challenge the dominant culture to live up to the lofty, liberating promises it makes in the name of promoting its new technologies. The tech-counterculture does not sit at the planning table; it blithely declares, rather: "Let's go ahead and behave as if these claims for liberation really were authentic!"

Radical Technologist

The second response, that of the radical technologist, depends more upon pragmatic than utopian knowledge and, in accepting advanced technology as a condition of possibility, it rejects the technophobia that is deeply entrenched in the tradition of left cultural despair. It describes those technological practices that are oppositional or alternative, and that are aimed at beating the military-industrial-media complex on its own ground, or in the now classic cyberpunk phrasing: "using technology before it is used on you." These practices range from low-level cybernetic sabotage in daily workplaces to the establishment of alternative media institutions that appropriate or utilize advanced technology for radical democratic ends; from workaday time stolen from an employer to independently funded institutions with the resources to rent time on a satellite transponder to beam TV signals.

    This spectrum runs from the smallest acts of resistance, under the worst possible circumstances, to the most creative political uses of existing conditions and technologies. There are few activist groups or advocacy publications that will not eagerly exploit the benefits of any new communications technology to come their way. And for every teen hacker driven by an inchoate desire to beat the system, there are a hundred information system and bulletin board users who support and maximize the extensive network of high-tech circuits of alternative information. The rules of thumb here are not governed by the ancient mysteries or by especially noble aspirations; they are survivalist rules, stripped of humanist pieties and leftist technophobia and driven instead by pragmatic know-how, by workable strategies and by local tactics.

The rules of action are equally unhampered by any moralistic aversion to using technology that has been developed and tested by the military. No one's hands are clean. The guiding assumption here, from the get-go, is that the new forms of smartness are almost certainly new forms of elite domination, and that we have to be better than we have been in the past in implementing these technologies in order to outsmart the forces of darkness.

Radical Ecologist

The practitioners of these guerilla arts and practices consciously appeal to well-established traditions of public communication—libertarian, republican-communitarian, anarchist—but their pragmatic concerns rarely extend to critical reexaminations of the ideologies of today's advanced technoculture. In this respect, it is the third and last response, the radical ecologist, that seems to have most to offer to the task of rethinking the place of technology in our culture. The broad ecology spectrum has its own utopian wing—both preindustrial and postindustrial—just as it boasts its own pragmatic movements, among them proponents of soft-energy paths, alternative energy systems, intermediate technologies, steady-state economies, and the like. But the deepest potential for change lies in the modification of cultural consciousness, and it is the increase in ecological awareness that underpins much of the impulse to attribute smartness to the object world.

As far as modern philosophies of nature go, the concept of the smart world is pretty far advanced. Chaos theory is a good example, in which natural processes are seen to be self-organizing, spontaneously generating patterns of order, stability, and diversity in situations once thought to be unstructured and chaotic, or dumb and inert. A good deal of ecological thought and science draws upon the premise of self-regulating ecosystems that can adapt through variation and diversification to environmental changes, even those changes that, through unsustainable use of resources, threaten the system with exhaustion.

In political and moral philosophy we have seen the growth of a movement to transfer rights associated with modern liberal societies onto nonhuman subjects and objects. The Endangered Species Act (1973) and various other legal decisions in the early 1970s opened the U.S. courts to the possibility of claims (contested daily by Reagan-Bush appointees) on behalf of environmental actors like rivers, beaches, and wetlands. The classic question remains the one raised by Christopher Stone's influential 1974 essay, *Should Trees Have Standing?*[1] Attributing ethical value to the land community or conferring actionable rights upon nonhuman agents seen as victims of environmental domination may be a reflection of the kind of ecological society we desire, but such arguments can only be made by human agents. They do not arise out of intrinsic or inalienable claims that lie outside of human society, within "nature" itself. Consequently, there is a thin line between acknowledging the sentience of the nonhuman natural world and assuming that it has an ethical capacity of its own. There is a slippery slope that runs from biocentric ethics, wherein all life forms are equal, to the diminution of our own hard-won social rights and freedoms, no sooner achieved than transferred elsewhere, to agents seen as more worthy because they are "closer" to nature.

Indeed, it is in the most extreme extensions of biocentric ethics, like the Gaia hypothesis, that we find the concept of the smart planet enjoying superior rights to those of its human occupants. Gaia, as theorized by James Lovelock, is that complex entity comprising the biosphere, the atmosphere, oceans, and land surface, which constantly seeks a stable physical and chemical environment for the planet. Life, as it evolved, was a necessary component to the homeostatic maintenance of the earth's properties. The biosphere therefore produced life species, one of which being human, to help it perform its work. Earth chauvinism rears its unlikely head when the development of a species threatens Gaia's needs. For the Gaian, the planetary organism, conceived as a whole, is smart enough to recognize that it is not in its best interest to tolerate the dumbness of the human species for much longer. That species will be eliminated (through global warming or whatever), because what matters most is the maintenance of Gaian life, not human life. On the one hand, this aspect of the Gaia hypothesis can be interpreted as a challenge to humanist philosophies of power that can still sanction species extinction. On the other hand, it can be seen as the flipside of such philosophies, the exact reverse of anthropocentrism, and thus a continuation of the logic of genocide by other means. Gaia's picture of a smart world demonstrates the clear danger of projecting human qualities onto nonhuman agents.

Smart New World

There is no question that today we need an alternative ideology of a smart world, an ideology of smartness that is not defined solely by the military-industrial–funded sociopaths at MIT, nor by the development brokers at the World Bank, nor by the free marketeers whose gospel is GATT; an ideology that is not exclusively tailored to the balance sheets of the resource-minded environmentalists nor governed by "ecofascist" fantasies about the primacy of Earth. Whatever the shape of this ideology, it will provide some of the rationale for our idea of a New World Order, whose historic birth should have been marked by the 1991 Earth Summit in Rio de Janeiro, in a year when the legacy of five centuries of colonial ecocide was under review. Few of us may have liked the sound of these words—New World Order—on the lips of George Bush, but it is clear that the moment of global politics is upon us, not only because environmental imperialism is one of the principal elements of geopolitics today, but also because ecological politics waged on a global scale is one of the few bulwarks left in the path of comprehensive capitalist control of the world economy.

[338]

Global agreements, like the "nonbinding principles" pursued, however ineffectually, in Rio are providing the only way for Southern nations to hold the Northern powers to ransom—the so-called negative power of the South—over matters like development aid and technology transfer. Despite the continuing toll exacted by the North on the Third World's natural resources, for the first time the South has had some kind of bargaining chip—its cooperation in global agreements about global warming and biodiversity and CFC use—which the North has little alternative but to buy off. One of the results has been to expose to public view the vicious spiral of (under) development and environmental degradation in a way that was never so visible in the neocolonial context of purely economic exploitation. Another was the sordid spectacle of how the South was bought off.

Sustainable Development

"Sustainable development," the guiding aegis of the so-called Earth Summit, has become an ideology in its own right. The World Commission on Environment and Development defines sustainability as development that "meets the needs of the present without compromising the ability of future generations to meet their own needs"; in the smart talk of futurological systems theory, this becomes a "society that has in place informational, social, and institutional mechanisms to keep in check the positive feedback loops that cause exponential population and capital growth."[2] Take your pick! Either formulation defines a smart, sustainable world in terms that are as far removed from the bio-centric worldview as from a definition of social and environmental justice adequate to the scale of global inequities.

Sustainable development was originally a set of development protocols applied to the use of foreign aid in Third World countries. The new need for universal cooperation on global agreements has, to some extent, shifted the focus of sustainability back upon the industrialized world, especially in the wake of the backlash generated by the hideous arrogance of Bush administration policy (an arrogance discreetly relished and welcomed by the other industrial nations in Rio). But one of the lessons of Rio was that a much broader debate has to take place about the value of what is being sustained, and the value of what is being claimed under the right to development if sustainable development is going to be anything more than a canny blueprint for the survival of global capitalism. Such a conversation, for example, would have to involve those indigenous peoples with no representation at the U.N.

and little more at the Earth Parliament in Rio, who are on the genocidal frontline of environmental exploitation in Northern and Southern nations alike. In matters of regulation, history and pragmatism show that there is little point in seeking a guarantee of environmental justice from the supervisory powers of U.N.-recognized nation-states, or from the World Bank–administered Global Environmental Facility, set up as a green fund for development. Global agreements brokered solely by the U.N. will barely curb the power of transnational corporations, defined loosely in the Rio agreements as benign agents with "rights," as if they were endangered species of dinosaurlike dimensions.

Indeed, one of the first acts of Dick Thornburgh, the Under-Secretary for Administration at the U.N., was to abolish the Department on Transnational Corporations, thereby eviscerating the code of conduct for transnationals that was being established at the department. Such acts were portents that the Rio agreements would be subject to the same fate as the New World Order formation of the Gulf War, similarly brokered in the name of the U.N. by the major industrialized powers. What has emerged from the summit is the blueprint for a world environmental market in the form of "smart" free-market solutions to problems of absorbing, distributing, and exploiting environmental costs. Within the U.S., the system of trading corporate pollution rights (complete with a central commodity auction administered by the good old Chicago Board of Trade) has been pioneered and partially legislated under the Bush administration's Clean Air Act (1991).[3]

The Rio agreements, especially the much amended Agenda 21, provide a structure for globalizing this market, from debt-for-nature swaps to the expedient bartering of individual states' regulatory controls in the interests of the free capital flow of international commerce. The logic of such a global market is perhaps the most gruesome comment on the explosive growth of the kind of corporate environmentalism that was brokered in Rio by the Business Council on Sustainable Development. Free marketeering, envisaged as the "natural," because self-regulating, economic solution to the crisis of nature.

For those who built the good intentions with which the road to the Summit (not the obligatory new road from the airport) were paved, the opportunity of Rio lay in its planetary address. For the first time, governments were officially responding to calls from the large environmental organizations and other NGOs (the broad category of nongovernment organizations that plays such a crucial role at the U.N.) for global thinking that was appropriate in scale and scope to the dimension of ecological crisis. For others, the globalism was the problem; the script was written for the big guys, and allowed all other nations, the NGOs, and other civic pressure groups only bit parts, all of which involved massive compromises at that. For them, a smart world order is one that recognizes signs of intelligent life in our own backyards, cities, and bioregions rather than in the universe at large, territorially controlled by the Security Council of the U.N. or by the movers and shakers at the World Bank. With globalism now on the table, there is no excuse for this kind of provincialism. Environmentally, smart policies in the Mississippi Delta mean something different in the Nile Delta or in the Amazon Basin. There's no global village, of course, but the villages are going global anyway.

Too Smart

The three responses I have briefly outlined each harbor some recognition that, in the current balance of technopower, it may be that some people, some things, and some ideas really are a little too smart for our good. They each acknowledge that it is up to us to recognize smartness and creativity and resistance in places, in people, and in things which the powers that be do not understand as smart; which they see as dumb, as obsolescent, as puny; which they see as ideas that don't count, and environments that can't last. The new smartness is an advanced form of competition in the sphere of intelligence, where knowledge, more than ever, is a species of power, and technology is its chief field of exercise.

Notes

James Der Derian

1. V. I. Lenin, *Selected Writings* (Moscow: Progress Publishers, 1977), p. 225.
2. Jean Baudrillard, *Simulations* (New York: Semiotext[e], 1983), p. 2.
3. Friedrich Nietzsche, *Twilight of the Idols* (Baltimore: Penguin, 1968), pp. 40–41; and James Der Derian, "Techno-Diplomacy," in *On Diplomacy: A Genealogy of Western Estrangement* (Oxford: Blackwell, 1987), pp. 199–200.
4. Baudrillard, *Simulations*, p. 48.
5. See Siegfried Kracauer, "Cult of Distraction: On Berlin's Picture Palaces," trans. T. Y. Levin in *New German Critique* 40 (Winter 1987), p. 95; and Kracauer's *Das Ornament*, trans. and ed. Levin (Cambridge, Mass.: Harvard University Press).
6. Walter Benjamin, "The Work of Art in the Age of Mechanical Reproduction," in *Illuminations*, ed. Hannah Arendt (New York: Schocken, 1969), pp. 241–242.
7. Guy Debord, *Society of the Spectacle* (Detroit: Black and Red, 1983), pp. 1, 23. In a more recent work, Debord persuasively—and somewhat despairingly—argues that the society of the spectacle retains its representational power in current times: see *Commentaires sur la société du spectacle* (Paris: Editions Gerard Lebovici, 1988).
8. Baudrillard, *Simulations*, p. 2. The original French version, *Simulacres et simulation* (Paris: Editions Galilee, 1981), has more on the simulacral nature of violence in cinema. In particular, see his readings of *China Syndrome, Barry Lyndon, Chinatown*, and *Apocalypse Now* (pp. 69–91).
9. Lenin, *Selected Writings*, p. 632.
10. E. David Cronon, ed., *The Political Thought of Woodrow Wilson* (Indianapolis: Bobbs-Merrill, 1965), pp. 482–484.
11. "Fatal Strategies," in *Jean Baudrillard: Selected Writings,* ed. Mark Poster (Stanford, Calif.: Stanford University Press, 1988), p. 191.
12. Whether it took the form of representing criminality on "America's Most Wanted," where alleged crimes are reenacted for the public benefit, or docudramatizing espionage on ABC prime-time news with a stand-in for the alleged spy Felix Bloch handing over a briefcase to a KGB stand-in, a genre of truthful simulations had already been established. There are the many commercially available war simulations, as well. To name a few: Harpoon, Das Boot Submarine, Wolf Pack, and Silent Service II (navy); Secret Weapons of the Luftwaffe, F-19 Stealth Fighter, A-10 Tank Killer, and F15 Strike Eagle (air force); and Populous, Balance of Power, SimCity, and Global Dilemma (more serious global simulations). On the heels of the Gulf War, wargames like Arabian Nightmare (in which the player has the option to kill American reporters like Ted Koppel) and the Butcher of Baghdad were added to the list.
13. This wargame differed from previous ones presented by Koppel (two on terrorism and one on nuclear war): there was not a pasha from Kissinger Associates in sight, and the talking heads barely had equal time with the video simulations. Constructed and narrated by the authors of the book *A Quick and Dirty Guide to War* and the wargame Arabian Nightmare, the program featured stock clips of war exercises, computer simulations of bombing runs, many maps, and a day-by-day, pull-down menu of escalating events. The postgame commentary (known in the ranks as a "hot wash-up") was conducted by two military analysts armed with pointers, James Blackwell and Harry Summers, Jr. They ended with a split decision—and a final cautionary note that "no plan survives contact with the enemy."
14. Simulations in this context could be broadly defined here as the continuation of war by means of verisimilitude and range from analytical games that use broad descriptions and a minimum of mathematical abstraction to make generalizations about the behavior of actors, to computerized models that use algorithms and high-resolution graphics to analyze and represent the amount of technical detail considered necessary to predict events and the behavior of actors.
15. Thomas B. Allen, *War Games: The Secret World of the Creators, Players, and Policy Makers Rehearsing World War III Today* (New York: McGraw-Hill, 1987), p. 4;

and "Nightline," 26 September 1990.
16. Two excellent criticisms of the internal assumptions of gaming can be found in a review of the literature by R. Ashley, "The Eye of Power: The Politics of World Modeling," *International Organization* 37, no. 3 (Summer 1983); and R. Hurwitz, "Strategic and Social Fictions in the Prisoner's Dilemma," in *International/Intertextual Relations*, ed. James Der Derian and Michael J. Shapiro (New York: Lexington Books, 1989), pp. 113–134.
17. Christopher M. Andrews, *KGB: The Inside Story of Its Operation from Lenin to Gorbachev* (New York: HarperCollins, 1991), pp. 583–605; and conversation with Oleg Gordievsky, 7–9 November 1991, Toronto, Canada.
18. The art of deterrence, prohibiting political war, favors the upsurge not of conflicts but *acts of war without war*. See Paul Virilio, *Pure War* (New York: Semiotext[e], 1983), p. 27.
19. Jean-Paul Sartre, "Vietnam: Imperialism and Genocide," in *Between Existentialism and Marxism* (New York: Pantheon, 1974), pp. 82–83.

Avital Ronell

1. Cited in Cathy Caruth, "Unclaimed Experience: Trauma and the Possibility of History," in *Literature and the E:thical Question*, ed. Claire Nouvet, special issue of *Yale French Studies* 79 (1991), p. 181.
2. This essay was written prior to the announcement of the jury's verdict of "not guilty." The logic of the ethical scream by which testimonial video is here understood appears to have stood its test: when the verdict of nonreading came out, the video produced effects of insurrection on the streets. It will not do to go with the unassailed diction of "burning down one's own neighborhood" when, in fact, it is a matter of radical expropriation and being-not-at-home which is the subject of this paper. Delivered in New York City, Oakland, Berkeley, and Los Angeles before and during the riots, this paper retains stylistic traces of the circumstances in which it was first presented. (I am not as opposed to

the usage of the word "riots" as others have been, no doubt because I am not seriously susceptible to supporting the order or logic of reason to which the riot addresses itself. As a type of noise and disturbance, "riot" belongs to the field of the ethical scream.) I have benefited greatly from the papers and discussions of Fred Moten, Peter Connor, Akira Lippit, and Shireen Patell.
3. Jacques Lacan, "Television," *October* 40 (Spring 1987), p. 36.
4. On interruption and the logic of destricturation, I follow Jacques Derrida's argument in his discussion of Levinas in "En ce moment même dans cet ouvrage me voici," in *Psyche: Inventions de l'autre* (Paris: Galilee, 1987), pp. 159–202.
5. A number of works have provided the frame within which to cast the catastrophic topicalities of television/video. I am assuming the reader's familiarity with the well-known essays by Mary Anne Doane, Meaghan Morris, John Hanhardt, Jonathan Crary, Patricia Mellencamp, Gilles Deleuze, and others. I would also like to acknowledge a paper by Deborah Esch, "No Time Like the Present: The 'Fact' of Television" presented at the University of California Humanities Research Institute, Irvine, in March 1992. It was here, while participating in the conference "Future Deconstructions" (Convener, Jonathan Culler), that I had the opportunity to follow the Rodney King trial in detail.
6. I am implicitly networking through the signifying chain of home-*homme* and homophobia. The technical media has thrown into relief an unreadable line between sexual marking and racial identification. On the law and its relation to sodomy in Freud, see my "The *sujet suppositaire*: Freud and the Rat Man," in *On Puns: The Foundations of Letters*, ed. Jonathan Culler (London: Blackwell, 1988), pp. 113–139, where I begin to develop a reading of the difference between detective and police work in terms of their relation to truth.
7. Television has made an increasing number of arrests in recent years; one show has recently celebrated its "two hundredth arrest."

8. See Jacques Derrida's remarkable discussion of Walter Benjamin's "Critique of Violence" in "Force of Law or the Mystical Foundations of Law," trans. Mary Quaintance, in *Cardozo Law Review* 2, pp. 5–6; *Deconstruction and the Possibility of Justice* (July–August 1990), p. 1009.

9. Ibid.

10. Cathy Caruth, "Introduction," in *American Imago* 48, no. 4, special issue, ed. Cathy Caruth; reprinted in *Psychoanalysis, Culture, Trauma* 11 (Winter 1991), p. 417. On the relation of atopy and flashback, Caruth writes that the "history of flashback tells—as psychiatry, psychoanalysis and neurobiology equally suggest—is, therefore, a history that literally *has no place*, neither in the past, in which it was not fully experienced, nor in the present, in which its precise images and enactments are not fully understood. In its repeated imposition as both image and amnesia, the trauma thus seems to evoke the difficult truth of a history that is constituted by the very incomprehensibility of its occurrence."

11. On an urgent inflection of this question, see also "'The AIDS Crisis Is Not Over': A Conversation with Gregg Bordowitz, Douglas Crimp, and Laura Pinsky," *American Imago* 48, no. 4, pp. 539–556.

12. Maurice Blanchot, "The Essential Solitude," in *The Space of Literature*, trans. Ann Smock (Lincoln: University of Nebraska Press, 1982), p. 32.

13. "Support Our Tropes" in *War After War* (San Francisco: City Lights Books, 1992), pp. 47–51. "Support Our Tropes II, or Why There Are So Many Cowboys in Cyburbia," *Yale Journal of Criticism* 5, no. 2 (Spring 1992), pp. 73–80.

14. Consider Paul de Man's demonstration of metonymic substitution in "Time and History in Wordsworth," *Diacritics* (Winter 1987), pp. 4–17.

15. For two relatively new readings of technicity and Ge-Stell in Heidegger I would suggest Veronique Foti's *Heidegger and the Poets: Poiesis/Sophia/Techné* (New Jersey: Humanities Press International, 1992), and especially Jean-Luc Nancy's title essay in *Une pensée finie* (Paris: Galilee, 1990).

16. Foti, *Heidegger and the Poets*, p. 4.

17. The tensional relationship between interpretation and reading is drawn from Andrzej Warminski, *Readings in Interpretation: Hölderlin, Hegel, Heidegger* (Minneapolis: University of Minnesota Press, 1987).

18. Sigmund Freud, *Beyond the Pleasure Principle*, in *The Standard Edition of the Complete Psychological Works of Sigmund Freud*, vol. 18, ed. James Strachey (London: Hogarth (1953–), p. 29 (translation modified).

19. Cathy Caruth, "Introduction," p. 7.

20. Kevin Newmark, "Traumatic Poetry: Charles Baudelaire and the Shock of Laughter," *American Imago* 48, no. 4, p. 534.

21. Shoshana Felman and Dori Laub, *Testimony: Crises of Witnessing in Literature, Psychoanalysis, and History* (New York: Routledge, 1991), p. 18.

22. Cited in B. A. van der Kolk and Onno van der Hart, "The Intrusive Past," *American Imago* 48, no. 4, p. 448.

23. Ibid., p. 449.

Kevin Robins

1. John Balzar, "Video Horror of Apache Victims' Deaths," *The Guardian*, 25 February 1991.

2. Lloyd deMause, "The Gulf War as Mental Disorder," *The Nation* 11 March 1991.

3. Ingrid Carlander, "Essor de la violence 'satanique' aux États-Unis," *Le monde diplomatique* (March 1990).

4. Susan Sontag, *On Photography* (Harmondsworth: Penguin, 1978), p. 20.

5. See *Cyborg Worlds*, ed. Les Levidow and Kevin Robins (London: Free Association Books, 1989).

6. Robert Jay Lifton and Eric Markusen, *The Genocidal Mentality* (London: Macmillan, 1990), pp. 106–107.

7. Ignacio Ramonet, "Télévision nécrophile," *Le monde diplomatique* (March 1990).

Les Levidow

1. Karl Marx, "The Fetishism of the Commodity and its Secret," in *Capital*, vol. 1 (New York: Penguin, 1976), pp. 164–166.
2. Herbert Marcuse, "Industrialization and Capitalism in the Work of Max Weber," in *Negations: Essays in Critical Theory* (Boston: Beacon, 1978), pp. 205, 217.
3. See Herbert Marcuse, *One Dimensional Man* (Boston: Beacon, 1964).
4. See Andrew Feenberg, "The Critical Theory of Technology," *Capitalism-Nature-Culture* 1, no. 5 (1990), pp. 17–45.
5. See Douglas N. Noble, "Mental Material," and Chris Gray, "The Cyborg Soldier," in *Cyborg Worlds: The Military Information Society*, ed. Les Levidow and Kevin Robins (London: Free Association Books, 1989), pp. 13–72. Also see Chris Gray, *Postmodern War: Computers as Weapons and Metaphors in the U.S. Military* (New York: Free Association Books, forthcoming).
6. *Digital Dialogues: Photography in the Age of Cyberspace*, special issue of *Ten.8* 2, no. 2 (Autumn 1991), pp. 56, 85.
7. Ibid., pp. 5, 61, 63.
8. George Lakoff, "War and Metaphor," *Propaganda Review* 8 (Fall 1991), pp. 18–21, 54–59.
9. See David Morrison and Paul Taylor, "Propaganda and the Media: The Gulf War," presented at Association for Cultural Studies Conference, 21–22 September 1991. Also see Bill MacGregor, "Death, Lies, and Videotape," *The Times Higher Educational Supplement* 10 January 1992, p. 13.
10. See *The Guardian*, 29 January 1991, p. 3.
11. See *Today*, 22 January 1991. Quoted in *The Guardian*, 23 January 1991.
12. See Mackenzie Wark, "War TV in the Gulf," *Meanjin* 50, no. 1 (Autumn 1991), p. 9.
13. See *Observer*, 20 January 1991.
14. Ibid.
15. *The Guardian*, 24 January 1991.
16. *Observer*, 20 January 1991.
17. Robert Stamm, "Mobilizing Fictions: The Gulf War, the Media, and the Recruitment of the Spectator," *Public Culture* 4, no. 2 (Spring 1992), pp. 101–126.
18. Ibid., p. 114.
19. See Les Levidow and Kevin Robins, "Vision Wars," *Race & Class* 32, no. 4 (April 1991), p. 325.
20. Douglas Kellner, *The Persian Gulf TV War* (Boulder: Westview Press, 1992), p. 247.
21. *Observer*, 20 January 1991.
22. Scott Montgomery, "Codes and Combat in Biomedical Discourse," *Science as Culture* 2, no. 3 (1991), pp. 341–390.
23. Asu Aksoy and Kevin Robins, "Exterminating Angels: Morality, Technology, and Violence in the Gulf War," *Science as Culture* 2, no. 3 (1991), pp. 322–337.
24. Tom Engelhardt, "The Gulf War as Total Television," *The Nation* 11 May 1992.
25. Gillian Skirrow, "Hellivision: An Analysis of Video Games," in *High Theory, Low Culture*, ed. Colin MacCabe (Manchester: Manchester University Press, 1986), p. 325.
26. Ibid., pp. 129–132.
27. Ibid., pp. 131–133. See Melanie Klein, *The Psycho-Analysis of Children* (London: Hogarth, 1975), pp. 182, 243–245.
28. John Broughton, "Babes in Arms: Object Relations and Fantasies of Annihilation," in *Psychology of War and Peace*, ed. R. Rieber (New York: Plenum, 1991), pp. 108–109.
29. Stamm, "Mobilizing Fictions," p. 104. See also Les Levidow and Kevin Robins, "The Eye of the Storm," *Screen* 32, no. 3 (Fall 1991), pp. 324–328.
30. See John Broughton, "The Pleasures of the Gulf War," in *Recent Trends in Theoretical Psychology*, vol. 3, ed. Robert Stamm (New York: Springer, 1993), pp. 231–246.
31. See Edward Said, *Orientalism: Western Conceptions of the Orient* (New York: Penguin, 1978), pp. 309–313.
32. *Time* 4 February 1991, pp. 12–13.
33. Kellner, *The Persian Gulf TV War*, pp. 297–309.
34. Michael Klare, "Behind Desert Storm: The New Military Paradigm," *Technology Review* (May–June, 1991), pp. 28–36.
35. Edward Said, "Empire of Sand," *Guardian Weekend Magazine* 12–13 January 1991.

36. Philip Agee, "Producing the Proper Crisis," *Z Magazine* 3, no. 11 (November 1990), p. 24.

37. Robins and Levidow, "Vision Wars."

38. See Stamm, "Mobilizing Fictions," p. 119.

39. See Zygmunt Bauman, *Modernity and the Holocaust* (Cambridge: Polity, 1989).

40. Stamm, "Mobilizing Fictions," p. 122.

41. Martha Rosler, "Image Simulations, Computer Manipulations," *Ten.8* 2, no. 2 (Autumn 1991), p. 63.

42. Levidow and Robins, *Cyborg Worlds*, p. 172.

43. See Timothy Druckrey, "Deadly Representations," *Ten.8* 2, no. 2 (Autumn 1991), pp. 22–23.

Andrew Ross

1. Christopher Stone, *Should Trees Have Standing? Towards Legal Rights for Natural Objects* (Los Altos, Calif.: W. Kaufman, 1974). For a broader account of this movement, see Roderick Nash, *The Rights of Nature: A History of Environmental Ethics* (Madison: University of Wisconsin Press, 1989).

2. This definition is taken from the sequel to the famous 1974 Club of Rome report: *Beyond the Limits: Confronting Global Collapse, Envisioning a Sustainable Future*, ed. Donella Meadows, Dennis L. Meadows, and Jorgen Randers (Post Mills, Vt.: Chelsea Green, 1992), p. 209.

3. See Brian Tokar, "Regulatory Sabotage," *Z Magazine* 5, no. 4 (April 1992), pp. 20–25.

Bibliography

ACT UP/NY Women's Book Group. *Women, AIDS, and Activism*. Boston: South End Press, 1992.

Allen, Thomas B. *War Games: The Secret World of the Creators, Players, and Policy Makers Rehearsing World War III Today*. New York: McGraw-Hill, 1987.

Anderson, Laurie. *Empty Places: A Performance*. New York: HarperPerennial, 1991.

Anderson, Laurie. *Stories from the Nerve Bible: A Retrospective, 1972-1992*. New York: HarperPerennial, 1994.

Anderson, Laurie. *United States*. New York: Harper & Row, 1984.

Anderson, Laurie. *Words in Reverse*. Buffalo, N.Y.: Hallwalls, 1979.

Aronowitz, Stanley. *The Crisis of Historical Materialism: Class, Politics, and Culture in Marxist Theory*. New York: Praeger, 1981.

Aronowitz, Stanley. *Dead Artists, Live Theories, and Other Cultural Problems*. New York: Routledge, 1994.

Aronowitz, Stanley. *False Promises: The Shaping of American Working Class Consciousness*. New York: McGraw-Hill, 1973.

Aronowitz, Stanley. *The Politics of Identity: Class, Culture, Social Movements*. New York: Routledge, 1992.

Aronowitz, Stanley. *Roll over Beethoven: The Return of Cultural Strife*. Hanover, N.H.: Wesleyan University Press, 1993.

Aronowitz, Stanley. *Science as Power: Discourse and Ideology in Modern Society*. Minneapolis: University of Minnesota Press, 1988.

Aronowitz, Stanley. *Working Class Hero: A New Strategy for Labor*. New York: Pilgrim Press, 1983.

Aronowitz, Stanley, and Henry A. Giroux. *Education under Siege: The Conservative, Liberal, and Radical Debate over Schooling*. South Hadley, Mass.: Bergin & Garvey, 1985.

Aronowitz, Stanley, and Henry A. Giroux. *Postmodern Education: Politics, Culture, and Social Criticism*. Minneapolis: University of Minnesota Press, 1991.

Attali, Jacques. *Millenium: Winners and Losers in the Coming World Order*. New York: Times Books, 1991.

Barrett, Edward, ed. *Sociomedia: Multimedia, Hypermedia, and the Social Construction of Knowledge*. Cambridge, Mass.: MIT Press, 1992.

Barrow, John D., *PI in the Sky: Counting, Thinking and Being*. New York: Oxford University Press, 1992.

Barry, John A. *Technobabble*. Cambridge, Mass.: MIT Press, 1991.

Bellin, David, and Gary Chapman, eds. *Computers in Battle: Will They Work?* Boston: Harcourt Brace Jovanovich, 1987.

Benedikt, Michael, ed. *Cyberspace: First Steps*. Cambridge, Mass.: MIT Press, 1991.

Benjamin, Marina, ed. *A Question of Identity: Women, Science, and Literature*. New Brunswick, N.J.: Rutgers University Press, 1993.

Benjamin, Walter. *Illuminations*. Edited by Hannah Arendt. New York: Shocken, 1968.

Berleur, Jacques, Andrew Clement, Richard Sizer, and Diane Whitehouse, eds. *The Information Society: Evolving Landscapes*. New York: Springer Verlag, 1990.

Bird, John, et al. *Mapping the Futures: Local Cultures, Global Change*. London: Routledge, 1993.

Blau, Herbert. *To All Appearances: Ideology and Performance*. New York: Routledge, 1992.

Boden, Margaret, ed. *The Philosophy of Artificial Intelligence*. New York: Oxford University Press, 1990.

Brecht, Bertolt. *Brecht on Theatre: The Development of an Aesthetic*. Edited and translated by John Willett. New York: Hill and Wang, 1964 [1982].

Broadbent, Donald, ed. *The Simulation of Human Intelligence*. Oxford: Blackwell, 1993.

Bukatman, Scott. *Terminal Identity: The Virtual Subject in Postmodern Science Fiction*. Durham, N.C.: Duke University Press, 1993.

Canguilhem, Georges. *The Normal and the Pathological*. Translated by Carolyn R. Fawcett in collaboration with Robert S. Cohen. New York: Zone Books, 1991.

Carroll, Noel E. *The Philosophy of Horror, or, Paradoxes of the Heart*. New York: Routledge, 1990.

Carter, Erica, and Simon Watney, eds. *Taking Liberties: AIDS and Cultural Politics*. London: Serpent's Tail, 1989.

Castells, Manuel. *The Information City: Information, Technology, Economic Restructuring, and the Urban Regional Process*. Oxford: Blackwell, 1989.

Caudill, Maureen. *In Our Own Image: Building an Artificial Person*. New York: Oxford University Press, 1992.

Chomsky, Noam. *Necessary Illusions: Thought Control in Democratic Societies*. Boston: South End Press, 1989.

Chomsky, Noam. *Year 501: The Conquest Continues*. Boston: South End Press, 1993.

Cirincione, Janine, and Brian D'Amato, eds. *Through the Looking Glass: Artists' First Encounters with Virtual Reality*. Exhibition catalogue. New York: Jack Tilton Gallery, 1992.

Collins, Jim. *Architecture of Excess: Cultural Life in the Information Age*. New York: Routledge, 1993.

Conley, Verena Andermatt, ed. *Rethinking Technologies*. Minneapolis: University of Minnesota Press, 1993.

Crandall, B. C., and James Lewis. *Nanotechnology: Research and Perspectives*. Cambridge, Mass.: MIT Press, 1992.

Crary, Jonathan. *Techniques of the Observer: On Vision and Modernity in the Nineteenth Century*. Cambridge, Mass.: MIT Press, 1990.

Crary, Jonathan, and Sanford Kwinter, eds. *Incorporations*. New York: Zone Books, 1992.

Critical Art Ensemble. *The Electronic Disturbance*. New York: Semiotext[e], 1994.

Cross, Brian. *It's Not about a Salary: Rap, Race, and Resistance in Los Angeles*. London: Verso, 1993.

Curran, James, and Michael Gurevitch, eds. *Mass Media and Society*. London: E. Arnold, 1991.

Danielson, Peter. *Artificial Morality: Virtuous Robots for Virtual Games*. London: Routledge, 1992.

Davies, Kath, Julienne Dickey, and Teresa Stratford. *Out of Focus: Writings on Women and the Media*. London: Women's Press, 1987.

De Landa, Manuel. *War in the Age of Intelligent Machines*. New York: Zone Books, 1991.

Der Derian, James. *Antidiplomacy: Spies, Terror, Speed, and War*. Cambridge, Mass.: Blackwell, 1992.

Der Derian, James. *On Diplomacy: A Genealogy of Western Estrangement*. Oxford: Blackwell, 1987.

Der Derian, James, and Michael J. Shapiro. *International/Intertextual Relations: Postmodern Readings of World Politics*. Lexington, Mass.: Lexington Books, 1989.

Dery, Mark, ed. *Flame Wars: The Discourse of Cyberculture*. Durham, N.C.: Duke University Press, 1993.

Doray, Bernard. *From Taylorism to Fordism.* London: Free Association Books, 1988.

Dorfman, Ariel. *The Empire's Old Clothes: What the Lone Ranger, Babar, and Other Innocent Heroes Do to Our Minds.* New York: Pantheon Books, 1983.

Dorfman, Ariel, and Armand Mattelart. *How to Read Donald Duck: Imperialist Ideology in the Disney Comic.* New York: International General, 1975.

Downing, John, Ali Mohammadi, and Annabelle Sreberny-Mohammadi. *Questioning the Media: A Critical Introduction.* Newbury Park, Calif.: Sage Publications, 1990.

Dreyfus, Hurbert. *What Computers Can't Do: A Critique of Artificial Reason.* New York: Harper & Row, 1972.

Dreyfus, Hurbert. *What Computers Still Can't Do.* Cambridge, Mass.: MIT Press, 1992.

Druckrey, Timothy, ed. *Iterations: The New Image.* Cambridge, Mass.: MIT Press, 1993.

Druckrey, Timothy, and Nadine Lemmon, eds. *For a Burning World is Come to Dance Inane: Essays by and about Jim Pomeroy.* Brooklyn, N.Y.: Critical Press, Inc., 1993.

Dyson, Frances, and Douglas Kahn. *Telesthesia.* Exhibition catalogue. San Francisco: San Francisco Art Institute, 1991.

Eccles, John C. *Evolution of the Brain: Creation of the Self.* London: Routledge, 1989.

Eldridge, John. *Getting the Message: News, Truth, and Power.* London: Routledge, 1993.

Ellul, Jacques. *The Technological Bluff.* Grand Rapids, Mich.: W. B. Eerdmans, 1990.

Felman, Shoshana, and Dori Laub. *Testimony: Crises of Witnessing in Literature, Psychoanalysis, and History.* New York: Routledge, 1991.

Firestone, Shulamith. *The Dialectic of Sex: The Case for Feminist Revolution.* New York: Bantam, 1971.

Fiske, John. *Power Plays, Power Works.* London: Routledge, 1993.

Forester, Tom, ed. *The Information Technology Revolution.* Cambridge, Mass.: MIT Press, 1985.

Forester, Tom, and Perry Morrison. *Computer Ethics: Cautionary Tales and Ethical Dilemmas in Computing.* Cambridge, Mass.: MIT Press, 1990.

Foti, Veronique. *Heidegger and the Poets: Poiesis/Sophia/Techné.* New Jersey: Humanities Press International, 1992.

Frank, Francine Wattman, and Paula A. Treichler. *Language, Gender, and Professional Writing: Theoretical Approaches and Guidelines for Nonsexist Usage.* New York: Commission on the Status of Women in the Profession, Modern Language Association of America, 1989.

Gandy, Oscar H. *The Panoptic Sort: A Political Economy of Personal Information.* Boulder, Colo.: Westview Press, 1993.

Garber, Marjorie B., Jann Matlock, and Rebecca Walkowitz. *Media Spectacles.* New York: Routledge, 1993.

Garson, Barbara. *The Electronic Sweatshop: How Computers Are Turning the Office of the Future into the Factory of the Past.* New York: Simon & Schuster, 1988.

Gill, Dawn, and Les Levidow, eds. *Anti-Racist Science Teaching.* London: Free Association Books, 1987.

Gooding-Williams, Robert, ed. *Reading Rodney King/Reading Urban Uprising.* New York: Routledge, 1993.

Göranzon, Bo, and Magnus Florin, eds. *Dialogue and Technology: Art and Knowledge.* London: Springer Verlag, 1991.

Gray, Ann. *Video Playtime: The Gendering of a Leisure Technology.* London: Routledge, 1992.

Grossberg, Lawrence, Cary Nelson, and Paula A. Treichler, eds. *Cultural Studies.* New York: Routledge, 1992.

Guattari, Félix. *Molecular Revolution: Psychiatry and Politics.* Translated by Rosemary Sheed. New York: Penguin, 1984.

Hafner, Katie, and John Markoff. *Cyberpunk: Outlaws and Hackers on the Computer Frontier*. New York: Touchstone, Simon & Schuster, 1991.

Hall, Stephen S. *Mapping the Next Millennium: The Discovery of New Geographies*. New York: Vintage Books, 1993.

Haraway, Donna. *Primate Visions: Gender, Race, and Nature in the World of Modern Science*. New York: Routledge, 1989.

Haraway, Donna. *Simians, Cyborgs, and Women: The Reinvention of Nature*. New York: Routledge, 1991.

Harding, Sandra G. *Feminism and Methodology: Social Science Issues*. Bloomington: Indiana University Press, 1987.

Harding, Sandra G., ed. *The "Racial" Economy of Science: Toward a Democratic Future*. Bloomington: Indiana University Press, 1993.

Harding, Sandra G. *The Science Question in Feminism*. Ithaca, N.Y.: Cornell University Press, 1986.

Harding, Sandra G. *Whose Science? Whose Knowledge? Thinking from Women's Lives*. Ithaca, N.Y.: Cornell University Press, 1991.

Harding, Sandra G., and Jean F. O'Barr, eds. *Sex and Scientific Inquiry*. Chicago: University of Chicago Press, 1987.

Hardison, O. B. *Disappearing through the Skylight: Culture and Technology in the Twentieth Century*. New York: Viking, 1989.

Hattinger, Gottfried, and Peter Weibel. *Out of Control: Ars Electronica 1991*. Linz: Landesverlag, 1991.

Hayward, Philip, ed. *Culture, Technology, and Creativity in the Late Twentieth Century*. London: J. Libbey, 1990.

Hebdige, Dick. *Cut 'n' Mix: Culture, Identity, and Caribbean Music*. London: Methuen, 1987.

Hecht, Frederick, and Joan Marks, eds. *Trends and Teaching in Clinical Genetics*. New York: A. R. Liss, 1977.

Heims, Steve J. *Constructing a Social Science for Postwar America: The Cybernetics Group 1946–1953*. Cambridge, Mass.: MIT Press, 1991.

Hendin, David, and Joan Marks. *The Genetic Connection: How to Protect Your Family against Hereditary Disease*. New York: Morrow, 1978.

Herdt, Gilbert, and Shirley Lindenbaum. *The Time of AIDS: Social Analysis, Theory, and Method*. Newbury Park, Calif.: Sage, 1992.

Hirsch, Eric, and Roger Silverstone, eds. *Consuming Technologies: Media and Information in Domestic Spaces*. London: Routledge, 1992.

Howell, John. *Laurie Anderson*. New York: Thunder's Mouth Press, 1992.

Hubbard, Ruth, and Elijah Wald. *Exploding the Gene Myth: How Genetic Information Is Produced and Manipulated by Scientists, Physicians, Employers, Insurance Companies, Educators, and Law Enforcers*. Boston: Beacon Press, 1993.

Huyssen, Andreas. *After the Great Divide: Modernism, Mass Culture, Postmodernism*. Bloomington: Indiana University Press, 1986.

Jameson, Fredric. *Postmodernism or the Cultural Logic of Late Capitalism*. Durham, N.C.: Duke University Press, 1991.

Jay, Martin. *Downcast Eyes: The Denigration of Vision in Twentieth-Century French Thought*. Berkeley: University of California Press, 1993.

Jay, Martin. *Force Fields: Between Intellectual History and Cultural Critique*. New York: Routledge, 1993.

Jenkins, Henry. *Textual Poachers: Television Fans and Participatory Culture*. New York: Routledge, 1992.

Jenks, Chris, ed. *Cultural Reproduction*. London: Routledge, 1993.

Johnson, George. *In the Palaces of Memory: How We Build the Worlds inside Our Heads*. New York: Knopf, 1991.

Johnson, George. *Machinery of the Mind: Inside the New Science of Artificial Intelligence*. New York: Times Books, 1986.

Jonas, Hans. *Philosophical Essays: From Ancient Creed to Technological Man*. Englewood Cliffs, N.J.: Prentice Hall, 1974.

Kadrey, Richard. *Covert Culture Sourcebook 2*. New York: St. Martin's Press, 1994.

Kahn, Douglas, and Gregory Whitehead, eds. *Wireless Imagination: Sound, Radio, and the Avant-Garde*. Cambridge, Mass.: MIT Press, 1992.

Kamper, Dietmar, and Christoph Wulf. *Looking Back on the End of the World*. New York: Semiotext[e], 1989.

Keller, Evelyn Fox. *Secrets of Life, Secrets of Death: Essays on Language, Gender, and Science*. New York: Routledge, 1992.

Kellner, Douglas. *Television and the Crisis of Democracy*. Boulder, Colo.: Westview Press, 1990.

Kelly, Janis, and Cindy Patton. *Making It: A Woman's Guide to Sex in the Age of AIDS*. Boston: Firebrand, 1987.

Kevles, Daniel J. *In the Name of Eugenics: Genetics and the Uses of Human Heredity*. New York: Knopf, 1985.

Kevles, Daniel J., and Leroy Hood, eds. *The Code of Codes: Scientific and Social Issues in the Human Genome Project*. Cambridge, Mass.: Harvard University Press, 1992.

Kidder, Tracy. *The Soul of a New Machine*. Boston: Little, Brown, 1981.

Kimbrell, Andrew. *The Human Body Shop: The Engineering and Marketing of Life*. San Francisco: Harper San Francisco, 1993.

Kittler, Friedrich. *Discourse Networks, 1800-1900*. Translated by Michael Metteer with Chris Cullens. Stanford, Calif.: Stanford University Press, 1990.

Klüver, Billy, Julie Martin, and Barbara Rose. *Pavilion*. New York: E. P. Dutton & Co., Inc., 1972.

Kramarae, Cheris, ed. *Technology and Women's Voices: Keeping in Touch*. London: Routledge & Kegan Paul, 1988.

Kramarae, Cheris, and Paula A. Treichler. *A Feminist Dictionary*. London: Pandora Press, 1985.

Kristeva, Julia. *Powers of Horror: An Essay on Abjection*. Translated by Leon S. Roudie. New York: Columbia University Press, 1982.

Kroker, Arthur. *SPASM: Virtual Reality, Android Music, and Electric Flesh*. New York: St. Martin's Press, 1993.

Kubey, Robert, and Mihaly Csikszentmihalyi. *Television and the Quality of Life: How Viewing Shapes Everyday Experience*. Hillsdale, N. J.: Lawrence Erlbaum Associates, 1990.

Kuhn, Annette, ed. *Alien Zone: Cultural Theory and Contemporary Science Fiction Cinema*. London: Verso, 1990.

Kuhns, William. *The Post-Industrial Prophets: Interpretations of Technology*. New York: Weybright & Talley, 1971.

Kurzweil, Raymond. *The Age of Intelligent Machines*. Cambridge, Mass.: MIT Press, 1990.

Landow, George P. *Hypertext: The Convergence of Contemporary Critical Theory and Technology*. Baltimore: Johns Hopkins University Press, 1992.

Lanham, Richard A. *The Electronic Word: Democracy, Technology, and the Arts*. Chicago: University of Chicago Press, 1993.

Lee, Martin A., and Norman Solomon. *Unreliable Sources: A Guide to Detecting Bias in the News Media*. New York: Carol Publishing Group, 1990.

Leebaert, Derek, ed. *The Future of Software*. Cambridge, Mass.: MIT Press, 1994.

Levidow, Les, ed. *Radical Science Essays*. London: Free Association Books, 1986.

Levidow, Les, ed. *Science as Politics*. London: Free Association Books, 1986.

Levidow, Les, and Bob Young. *Science, Technology and the Labour Process*. London: CSE Books; Atlantic Highlands, N.J.: Humanities Press, 1981.

Levidow, Les, and Kevin Robins, eds. *Cyborg Worlds: The Military Information Society*. London: Free Association Books, 1989.

Levy, Steven. *Artificial Life: The Quest for a New Creation*. New York: Vintage Books, 1992.

Lewis, H. W. *Technological Risk*. New York: W. W. Norton, 1990.

Lewis, Justin. *The Ideological Octopus: An Exploration of Television and Its Audience*. New York: Routledge, 1991.

Lewontin, R. C. *Biology as Ideology: The Doctrine of DNA*. New York: HarperCollins, 1991.

Lewontin, R. C. *The Genetic Basis of Evolutionary Change*. New York: Columbia University Press, 1974.

Lewontin, R. C., Steven Rose, and Leon J. Kamin. *Not in Our Genes: Biology, Ideology, and Human Nature*. New York: Pantheon Books, 1984.

Lovejoy, Margot. *Postmodern Currents: Art and Artists in the Age of Electronic Media*. Englewood Cliffs, N.J.: Prentice Hall, 1992.

Lubar, Steven D. *Infoculture: The Smithsonian Book of Information Age Inventions*. Boston: Houghton Mifflin Co., 1993.

Lury, Celia. *Cultural Rights: Technology, Legality, and Personality*. London: Routledge, 1993.

Lyon, David. *The Electronic Eye: The Rise of Surveillance Society*. Minneapolis: University of Minnesota Press, 1994.

MacArthur, John R. *Second Front: Censorship and Propaganda in the Gulf War*. New York: Hill and Wang, 1992.

MacCabe, Colin, ed. *High Theory/Low Culture: Analysing Popular Television and Film*. Manchester: Manchester University Press, 1986.

Malcolm, Janet. *The Purloined Clinic: Selected Writings*. New York: Vintage Books, 1993.

Marks, Joan H., ed. *Advocacy in Health Care: The Power of a Silent Constituency*. Clifton, N.J.: Humana Press, 1986.

Marks, Joan H., et al. *Genetic Counseling Principles in Action: A Casebook*. White Plains, N.Y.: The Foundation, 1989.

Massey, Doreen, Paul Quintas, and David Wield, eds. *High Tech Fantasies: Science Parks in Society, Science, and Space*. London: Routledge, 1992.

Mattelart, Armand. *Advertising International: The Privatization of Public Space*. London: Routledge, 1991.

Mattelart, Armand. *Mass Media, Ideologies, and the Revolutionary Movement*. Sussex, England: Harvester Press, 1980.

Mattelart, Armand. *Multinational Corporations and the Control of Culture: The Ideological Apparatuses of Imperialism*. Sussex, England: Harvester Press, 1979.

Mattelart, Armand. *Transnationals and the Third World: The Struggle for Culture*. Translated by David Buxton. South Hadley, Mass.: Bergin & Garvey, 1983.

Mattelart, Armand, and Hector Schmucler. *Communication and Information Technologies: Freedom of Choice for Latin America?* Translated by David Buxton. Norwood, N.J.: Ablex Publishing Corp., 1985.

Mattelart, Armand, and Michèle Mattelart. *Rethinking Media Theory: Signposts and New Directions*. Minneapolis: University of Minnesota Press, 1992.

Mattelart, Armand, and Seth Siegelaub, eds. *Communication and Class Struggle: An Anthology in 2 Volumes*. New York: International General, 1979–1983.

Mattelart, Armand, Xavier Delcourt, and Michele Mattelart. *International Image Markets: In Search of an Alternative Perspective*. Translated by David Buxton. London: Comedia Publishing Group, 1984.

Mattelart, Michèle. *Women, Media, and Crisis: Femininity and Disorder.* London: Comedia Publishing Group, 1986.

Mattelart, Michèle, and Armand Mattelart. *The Carnival of Images: Brazilian Television Fiction.* New York: Bergin & Garvey, 1990.

McLuhan, Marshall. *Understanding Media: The Extensions of Man.* New York: New American Library, 1964.

Miller, Mark Crispin. *Boxed In: The Culture of TV.* Evanston, Ill.: Northwestern University Press, 1988.

Mitchell, William J. *The Reconfigured Eye: Visual Truth in the Post-Photographic Era.* Cambridge, Mass.: MIT Press, 1991.

Moravec, Hans P. *Mind Children: The Future of Robot and Human Intelligence.* Cambridge, Mass.: Harvard University Press, 1988.

Mowlana, Hamid, George Gerbner, and Herbert I. Schiller, eds. *Triumph of the Image: The Media's War in the Persian Gulf—A Global Perspective.* Boulder, Colo.: Westview Press, 1992.

Munitz, Milton Karl. *The Question of Reality.* Princeton, N.J.: Princeton University Press, 1990.

Nash, Roderick. *The Rights of Nature: A History of Environmental Ethics.* Madison Wis.: University of Wisconsin Press, 1989.

Nye, David E. *American Technological Sublime.* Cambridge, Mass.: MIT Press, 1994.

O'Malley, Padraig, ed. *The AIDS Epidemic: Private Rights and the Public Interest.* Boston: Beacon Press, 1988.

Olalquiaga, Celeste. *Megalopolis: Contemporary Cultural Sensibilities.* Minneapolis: University of Minnesota Press, 1992.

Penley, Constance, ed. *Feminism and Film Theory.* New York: Routledge, 1988.

Penley, Constance. *The Future of an Illusion: Film, Feminism, and Psychoanalysis.* Minneapolis: University of Minnesota Press, 1989.

Penley, Constance, et al., eds. *Close Encounters: Film, Feminism, and Science Fiction.* Minneapolis: University of Minnesota Press, 1991.

Penley, Constance, and Andrew Ross, eds. *Technoculture.* Minneapolis: University of Minnesota Press, 1991.

Penley, Constance, and Sharon Willis, eds. *Male Trouble.* Minneapolis: University of Minnesota Press, 1993.

Pomeroy, Jim. *Apollo Jest: An American Mythology (in depth).* San Francisco: Blind Snake Blues, 1983.

Poster, Mark. *The Mode of Information: Poststructuralism and Social Context.* Chicago: University of Chicago Press, 1990.

Postman, Neil. *Technopoly: The Surrender of Culture to Technology.* New York: Knopf, 1992.

Poulain, Jacques, and Wolfgang Schirmacher. *Penser après Heidegger: Actes du colloque du centenaire, Paris, 25–27 Septembre 1989.* Paris: L'Harmattan, 1992.

Preston, William Jr., Edward S. Herman, and Herbert I. Schiller. *Hope and Folly: The United States and UNESCO, 1945–1985.* Minneapolis: University of Minnesota Press, 1989.

Restak, Richard M. *Pre-Meditated Man: Bioethics and the Control of Future Human Life.* New York: Basic Books, 1977.

Rheingold, Howard. *Virtual Reality: The Revolutionary Technology of Computer-Generated Artificial Worlds and How It Promises to Transform Society.* New York: Simon & Schuster, 1992.

Ritchin, Fred. *In Our Own Image: The Coming Revolution in Photography.* New York: Aperture, 1990.

Robins, Kevin, and Frank Webster. *The Technical Fix: Education, Computers, and Industry.* New York: St. Martin's Press, 1989.

Ronell, Avital. *Crack Wars: Literature, Addiction, Mania.* Lincoln: University of Nebraska Press, 1992.

[353]

Ronell, Avital. *Dictations: On Haunted Writing*. Bloomington: Indiana University Press, 1986.

Ronell, Avital. *The Telephone Book: Technology—Schizophrenia—Electric Speech*. Lincoln: University of Nebraska Press, 1989.

Rose, Tricia. *Black Noise: Rap Music and Black Culture in Contemporary America*. Hanover, N.H.: Wesleyan University Press, 1994.

Rose, Tricia, and Andrew Ross, eds. *Microphone Fiends: Youth Music and Youth Culture*. New York: Routledge, 1994.

Ross, Andrew. *The Failure of Modernism: Symptoms of American Poetry*. New York: Columbia University Press, 1986.

Ross, Andrew. *No Respect: Intellectuals and Popular Culture*. New York: Routledge, 1989.

Ross, Andrew. *Strange Weather: Culture, Science, and Technology in the Age of Limits*. London: Verso, 1991.

Ross, Andrew, ed. *Universal Abandon?: The Politics Of Postmodernism*. Minneapolis: University of Minnesota Press, 1988.

Rucker, Rudy B., R. U. Sirius, and Queen Mu. *Mondo: A User's Guide to the New Edge*. New York: HarperPerennial, 1992.

Ryan, Charlotte. *Prime Time Activism: Media Strategies for Grassroots Organizing*. Boston: South End Press, 1991.

Scarry, Elaine. *The Body in Pain: The Making and Unmaking of the World*. New York: Oxford University Press, 1985.

Scarry, Elaine, ed. *Literature and the Body: Essays on Populations and Persons*. Baltimore: John Hopkins University Press, 1988.

Schiller, Herbert I. *Communication and Cultural Domination*. White Plains, N.Y.: International Arts and Sciences Press, 1976.

Schiller, Herbert I. *Culture, Inc.: The Corporate Takeover of Public Expression*. New York: Oxford University Press, 1989.

Schiller, Herbert I. *Information and the Crisis Economy*. Norwood, N.J.: Ablex Publishing Corp., 1984.

Schiller, Herbert I. *Mass Communications and American Empire*. Boulder, Colo.: Westview Press, 1992.

Schiller, Herbert I. *The Mind Managers*. Boston: Beacon Press, 1973.

Schiller, Herbert I. *Who Knows: Information in the Age of the Fortune 500*. Norwood, N.J.: Ablex Publishing Corp., 1981.

Schiller, Herbert I., and Joseph D. Phillips, eds. *Super-State: Readings in the Military-Industrial Complex*. Urbana: University of Illinois Press, 1970.

Schiller, Herbert I., and Kaarle Nordenstreng, eds. *National Sovereignty and International Communication*. Norwood, N.J.: Ablex Publishing Co., 1979.

Schirmacher, Wolfgang. *Schopenhauers aktualität: Ein Philosoph wird neu gelesen / herausgegeben*. Vienna: Passagen, 1988.

Schirmacher, Wolfgang. *Technik und Gelassenheit: Zeitkritik nach Heidegger*. Freiburg: Alber, 1983.

Searle, John R. *The Rediscovery of the Mind*. Cambridge, Mass.: MIT Press, 1992.

Seiter, Ellen, ed. *Remote Control: Television, Audiences, and Cultural Power*. London: Routledge, 1989.

Seltzer, Mark. *Bodies and Machines*. New York: Routledge, 1992.

Sheehan, James J., and Morton Sosna, eds. *The Boundaries of Humanity: Humans, Animals, Machines*. Berkeley: University of California Press, 1991.

Silverstone, Roger, ed. *Consuming Technologies: Media and Information in Domestic Spaces*. London: Routledge, 1992.

Skovmand, Michael, and Kim Christian, eds. *Media Cultures: Reappraising Transnational Media*. London: Routledge, 1992.

Slobodkin, Lawrence B. *Simplicity and Complexity in Games of the Intellect*. Cambridge, Mass.: Harvard University Press, 1992.

Solomon, William S., and Robert W. McChesney. *Ruthless Criticism: New Perspectives in U.S. Communication History*. Minneapolis: University of Minnesota Press, 1993.

Solomonides, Tony, and Les Levidow, eds. *Compulsive Technology: Computers as Culture*. London: Free Association Books, 1985.

Spicker, Stuart F., Kathleen Woodward, and David D. Van Tassel, eds. *Aging and the Elderly: Humanistic Perspectives in Gerontology*. Atlantic Highlands, N.J.: Humanities Press, 1978.

Stafford, Barbara Maria. *Body Criticism: Imaging the Unseen in Enlightenment Art and Medicine*. Cambridge, Mass.: MIT Press, 1991.

Sterling, Bruce. *The Hacker Crackdown: Law and Disorder on the Electronic Frontier*. New York: Bantam Books, 1992.

Strathern, Marilyn. *Reproducing the Future: Anthropology, Kinship, and the New Reproductive Technologies*. New York: Routledge, 1992.

Taussig, Michael. *Mimesis and Alterity: A Particular History of the Senses*. New York: Routledge, 1993.

Taussig, Michael. *The Nervous System*. New York: Routledge, 1992.

Thagard, Paul. *Conceptual Revolutions*. Princeton, N.J.: Princeton University Press, 1992.

Treichler, Paula A., Cheris Kramarae, and Beth Stafford, eds. *For Alma Mater: Theory and Practice in Feminist Scholarship*. Urbana: University of Illinois Press, 1985.

Tuana, Nancy, ed. *Feminism and Science*. Bloomington: Indiana University Press, 1989.

Varela, Francisco J., Evan Thompson, and Eleanor Rosch. *The Embodied Mind: Cognitive Science and Human Experience*. Cambridge, Mass.: MIT Press, 1991.

Vattimo, Gianni. *The Transparent Society*. Baltimore: Johns Hopkins University Press, 1992.

Virilio, Paul. *The Aesthetics of Disappearance*. Translated by Philip Beitchman. New York: Semiotext[e], 1991.

Virilio, Paul. *The Lost Dimension*. Translated by Daniel Moshenberg. New York: Semiotext[e], 1991.

Waldrop, M. Mitchell. *Complexity: The Emerging Science at the Edge of Order and Chaos*. New York: Simon & Schuster, 1992.

Ward, Dean, and John Morgenthaler. *Smart Drugs and Nutrients: How to Improve Your Memory and Increase Your Intelligence Using the Latest Discoveries in Neuroscience*. Santa Cruz, Calif.: B&J, 1990.

Weizenbaum, Joseph. *Computer Power and Human Reason: From Judgement to Calculation*. San Francisco: W. H. Freeman, 1976.

Wheale, Peter, and Ruth M. McNally. *Genetic Engineering: Catastrophe or Utopia?* New York: St. Martin's Press, 1988.

White, Mimi. *Tele-Advising: Therapeutic Discourse in American Television*. Chapel Hill: University of North Carolina Press, 1992.

Winner, Langdon. *Autonomous Technology: Technics-Out-of-Control as a Theme in Political Thought*. Cambridge, Mass.: MIT Press, 1977.

Winner, Langdon, ed. *Democracy in a Technological Society*. Boston: Kluwer, 1992.

Winner, Langdon. *Techné and Politeia: The Technical Constitution of Society*. Los Angeles: School of Architecture and Urban Planning, University of California, 1981.

Winner, Langdon. *The Whale and the Reactor: A Search for Limits in an Age of High Technology*. Chicago: University of Chicago Press, 1986.

Winograd, Terry, and Fernando Flores. *Understanding Computers and Cognition: A New Foundation for Design*. Norwood, N.J.: Ablex Publishing Corp., 1986.

Woodward, Kathleen. *Aging and Its Discontents: Freud and Other Fictions*. Bloomington: Indiana University Press, 1991.

Woodward, Kathleen, ed. *Myths of Information: Technology and Postindustrial Culture*. Madison, Wis.: Coda Press, 1980.

Woodward, Kathleen, and Murray M. Schwartz, eds. *Memory and Desire: Aging—Literature—Psychoanalysis*. Bloomington: Indiana University Press, 1986.

Woodward, Kathleen, Teresa de Lauretis, and Andreas Huyssen, eds. *The Technological Imagination: Theories and Fictions*. Madison, Wis.: Coda Press, 1980.

Woolley, Benjamin. *Virtual Worlds: A Journey in Hype and Hyperreality*. Cambridge, Mass.: Blackwell Publishers, 1992.

Wright, Will. *Wild Knowledge: Science, Language, and Social Life in a Fragile Environment*. Minneapolis: University of Minnesota Press, 1992.

Zimmerman, Michael E. *Heidegger's Confrontation with Modernity: Technology, Politics, and Art*. Bloomington: Indiana University Press, 1990.

Zuboff, Shoshana. *In the Age of the Smart Machine: The Future of Work and Power*. New York: Basic Books, 1988.

Notes on Contributors

Laurie Anderson is a musician, performance artist, and videomaker. Her works, *The United States, Big Science, Home of the Brave,* and *Voices from the Beyond,* among others, have established a clear connection between emerging technology and the issues of human experience.

Stanley Aronowitz is Professor of Sociology at the Graduate Center of the City University of New York. He is the author of *Science as Power, The Crisis of Historical Materialism, Working Class Hero,* and *The Politics of Identity.* He is a member of the editorial collective of *Social Text,* convenor of the Union of Democratic Intellectuals, and a frequent speaker on issues of technology and science policy. His most recent book is *Dead Artists, Live Theories.*

Gretchen Bender is an artist who takes to task the proliferation of mass media. Her critical examination of the systems and strategies that underlie the mediated, televised, and filmic image offers a uniquely apocalyptic vision of communication in the late twentieth century. Exhibitions of her work and performances of her electronic media theater works have included presentations at the Moderna Museet, Stockholm, the Setagaya Museum, Tokyo, and The Museum of Modern Art, San Francisco.

Gary Chapman is Coordinator of the National Forum on Science and Technology at the LBJ School of Public Affairs, University of Texas, Austin. He is also Director of the 21st Century Project, a public-interest program exploring new directions for science and technology policy in the post-Cold War era. He is the coeditor of *Computers in Battle: Will They Work?* His articles have appeared in *Technology Review, Computer World, South Atlantic Quarterly,* and numerous other publications.

James Der Derian is Professor of Political Science at the University of Massachusetts, Amherst. His most recent book is *Antidiplomacy: Spies, Terror, Speed, and War,* and he is currently writing a book on war games, the media, and U.S. foreign policy entitled *Virtual Security.*

Timothy Druckrey is an independent curator, critic, and writer concerned with issues of photographic history, representation, and technology. He has taught in graduate programs at the International Center of Photography, the School of Visual Arts, and the Center of Creative Imaging. He co-organized the exhibition *Iterations: The New Image* at the International Center of Photography and edited the book published by MIT Press. He has contributed extensively to numerous publications, including *Afterimage*, *Aperture*, and *Views*. He is currently writing a study of the relationships between technology and photography called *Representation and Photography* (forthcoming from Manchester University Press).

Billy Klüver is an engineer who cofounded E.A.T. (Experiments in Art and Technology, Inc.) with Robert Rauschenberg in 1966. E.A.T., a not-for-profit service organization for artists and engineers, established a Technical Services Program to provide artists with technical information and assistance by matching them with engineers and scientists who collaborate with them. In addition, E.A.T. initiates and administers interdisciplinary projects involving artists with new technology. Klüver has lectured widely and has published numerous articles on art and technology. He is coauthor (with Julie Martin) of *Kiki's Paris*, coeditor of *Pavilion*, and the author of *A Day with Picasso*. Currently he is working with Julie Martin on a social history of the art community from 1945 to 1965.

Les Levidow has been Managing Editor of *Science as Culture* since its inception in 1987, and of its predecessor, the *Radical Science Journal*. He is coeditor of several books, including *Anti-Racist Science Teaching* and *Cyborg Worlds: The Military Information Society*. As Research Fellow at the Open University, he has been studying the safety regulation of agricultural biotechnology, colloquially known as "the real Jurassic Park." He has been actively involved in the Palestine Solidarity Campaign, as well as Return, a Jewish anti-Zionist group.

R.C. Lewontin is Alexander Agassiz Professor of Zoology and Professor of Biology at Harvard University. He is the author of *The Genetic Basis of Evolutionary Change* and *Biology as Ideology: The Doctrine of DNA*.

Joan H. Marks is Director of the Human Genetics Graduate Program at Sarah Lawrence College in Bronxville, New York. She is the author of several books on health care and a member of the advisory board of the Women's Health Initiative of the National Institutes of Health and the American Board of Internal Medicine.

Margaret Morse is Assistant Professor at the University of California, Santa Cruz, where she teaches film, television, video, and the criticism of electronic culture. She is currently working on two books, *Television Reality: Discursive Formats* and *Genres of Everyday Television and Virtualities.*

Simon Penny is Associate Professor of Art and Robotics at Carnegie Mellon University, Pittsburgh. Along with teaching in Electronic Media Arts, his recent activities include editing *Critical Issues in Electronic Media.* In 1993 he organized *Machine Culture*, an international survey exhibition of interactive media art at Siggraph '93. He makes interactive installations and mobile robotic artworks.

Kevin Robins is Reader in Cultural Geography at the Centre for Urban and Regional Development Studies, University of Newcastle upon Tyne. He is the author of *Geografia dei Media* and coeditor of *The Regions, the Nations, and the BBC.*

Avital Ronell is Professor of Theory and Comparative Literature at the University of California, Berkeley. She is the author of *The Telephone Book: Technology—Schizophrenia—Electric Speech, Crack Wars, Finitude's Score: Essays for the End of the Millenium, and Dictations: On Haunted Writing.*

Tricia Rose is Assistant Professor of History and Africana Studies at New York University. She is the author of *Black Noise: Rap Music and Black Culture in Contemporary America* and coeditor of *Microphone Fiends: Youth Music and Youth Culture* with Andrew Ross. Her essays on cultural politics, black popular music and culture, and popular cultural theory have appeared in the *Village Voice, Camera Obscura,* and the collection *Black Popular Culture.*

Andrew Ross is Director of American Studies at New York University and a columnist for *Artforum*. Recently he coedited *Microphone Fiends: Youth Music and Youth Culture* with Tricia Rose. His latest book is *The Chicago Gangster Theory of Life: Ecology, Culture, and Society*. He is also the author of *Strange Weather: Culture, Science, and Technology in the Age of Limits* and *No Respect: Intellectuals and Popular Culture*.

Elaine Scarry is Professor of English at Harvard University. She has written *The Body in Pain: The Making and Unmaking of the World* and is now completing a book about war and the social contract entitled *The Matter of Consent*. Her essays about the way the material world resists language are collected in *Resisting Representation*.

Herbert I. Schiller is Professor Emeritus of Communications at the University of California, San Diego. He is the author of numerous books which examine and critique the American informational condition. Among them are *Mass Communications and American Empire*, *Information and the Crisis Economy*, and *Culture, Inc.: The Corporate Takeover of Public Expression*.

Wolfgang Schirmacher is a philosopher of technology who has published several books in German and numerous articles on bioethics, environmental philosophy, and computer ethics. He currently teaches Media Philosophy at the New School for Social Research in New York and is Professor in the Science-Technology-Society Program at Penn State University. He serves also as president of the International Schopenhauer Association and is editor of *Schopenhauer Studien*.

Paula A. Treichler is on the faculty of the College of Medicine, the Institute of Communications Research, and the Women's Studies Program at the University of Illinois, Urbana-Champaign. She is co-author of *A Feminist Dictionary, Language, Gender, and Professional Writing: Theoretical Approaches and Guidelines for Nonsexist Usage*, and coeditor of *For Alma Mater: Theory and Practice in Feminist Scholarship*. Her work on AIDS has appeared in many journals, including *Cultural Studies*, *Artforum*, and *Transition*. She is completing a book on AIDS and culture.

Langdon Winner is Professor of Political Science in the Department of Science and Technology Studies at Rensselaer Polytechnic Institute. He is a political theorist who focuses upon social issues that surround modern technological change. He is the author of *Autonomous Technology, The Whale and the Reactor: A Search for Limits in an Age of High Technology*, and editor of *Democracy in a Technological Society*. Mr. Winner has lectured widely in the U.S. and Europe. His column "The Culture of Technology" appears regularly in *Technology Review*.

Kathleen Woodward is Director of the Center for Twentieth Century Studies, a postdoctoral research institute at the University of Wisconsin-Milwaukee, where she is a Professor of English. She has published essays on art and technology and on cybernetics and post-modern narrative, and is the editor of *The Myths of Information: Technology and Postindustrial Culture*, as well as coeditor (with Teresa de Lauretis and Andreas Huyssen) of *The Technological Imagination: Theories and Fictions*. Recently awarded a grant from the Rockefeller Foundation to establish a fellowship program in Age Studies, she is the author of *Aging and Its Discontents: Freud and Other Fictions*. She is currently at work on a book entitled *Mixed Emotions*.

DISCUSSIONS IN CONTEMPORARY CULTURE is an award–winning series of books copublished by Dia Center for the Arts, New York City, and Bay Press, Seattle. These volumes offer rich discourses on a broad range of cultural issues in formats that encourage scrutiny of diverse critical approaches and positions.

DISCUSSIONS IN CONTEMPORARY CULTURE
Edited by Hal Foster

VISION AND VISUALITY
Edited by Hal Foster

THE WORK OF ANDY WARHOL
Edited by Gary Garrels

REMAKING HISTORY
Edited by Barbara Kruger and Philomena Mariani

DEMOCRACY
A Project by Group Material
Edited by Brian Wallis

IF YOU LIVED HERE
The City in Art, Theory, and Social Activism
A Project by Martha Rosler
Edited by Brian Wallis

CRITICAL FICTIONS
The Politics of Imaginative Writing
Edited by Philomena Mariani

BLACK POPULAR CULTURE
A Project by Michele Wallace
Edited by Gina Dent

For information:
Bay Press
115 West Denny Way
Seattle, WA 98119
tel. 206.284.5913
fax. 206.284.1218